Remote Sensing Imagery for Natural Resources Monitoring

Methods and Cases in Conservation Science
Mary C. Pearl, Editor

Methods and Cases in Conservation Science

Thomas K. Rudel and Bruce Horowitz, *Tropical Deforestation: Small Farmers and Land Clearing in the Ecuadorian Amazon*

Joel Berger and Carol Cunningham, *Bison: Mating and Conservation in Small Populations*

Jonathan D. Ballou, Michael Gilpin, and Thomas J. Foose, *Population Management for Survival and Recovery: Analytical Methods and Strategies in Small Population Conservation*

Susan K. Jacobson, *Conserving Wildlife: International Education and Communication Approaches*

Gordon MacMillan, *At the End of the Rainbow? Gold, Land, and People in the Brazilian Amazon*

Perspectives in Biological Diversity Series

Holmes Rolston, III, *Conserving Natural Value*

Series Editor, Mary C. Pearl
Series Advisers, Christine Padoch and Douglas Daly

Remote Sensing Imagery for Natural Resources Monitoring

A Guide for First-Time Users

David S. Wilkie

and

John T. Finn

New York Columbia University Press

Columbia University Press
New York Chichester, West Sussex

Library of Congress Cataloging-in-Publication Data

Wilkie, David S.
Remote sensing imagery for natural resources monitoring : a guide for first-time users / David S. Wilkie and John T. Finn.
p. cm. — (Methods and cases in conservation science)
Includes bibliographical references and index.
ISBN 0-231-07928-1 (cl. : alk. paper). — ISBN 0-231-07929-X (pbk. alk. paper)
1. Remote sensing. I. Finn, John T. II. Series.
G70.4.W438 1996
621.36′78—dc20 95-22750

♾
Casebound editions of Columbia University Press books are printed on permanent and durable acid-free paper.

Printed in the United States of America

c 10 9 8 7 6 5 4 3 2 1
p 10 9 8 7 6 5 4 3 2 1

To Gilda and Elizabeth

Contents

Illustrations

Plates follow p. 138

Tables

Acknowledgments

We would like to thank Mary Pearl, the editor of the series, for giving us the opportunity to write this book and Connie Barlow for remarkably comprehensive and sage editorial advice. The last chapter was made possible by the authors of the papers that we use as case studies; all very graciously agreed to be interviewed, thus allowing us to take advantage of their collective experience. Special thanks, therefore, to John Sidle, Cheryl Pearce, Ian Barton, Jim Tucker, Mary Altalo, Steve Sader, Stan Herwitz, Tom Stone, and Mike Scott. An anonymous reviewer provided several helpful suggestions. Anne McCoy and Teresa Bonner of Columbia University Press provided clear and professional advice in dealing with our multiple graphics formats and helped enormously in navigating us through the revision and production process.

Remote Sensing Imagery for Natural Resources Monitoring

1

Purpose and Scope

The Sale of Human Land Transformation
Summary of Chapters

A quick look in *Books in Print* under the heading Remote Sensing will turn up a score of books on image processing, analysis, and interpretation. So why did we believe it worthwhile or even necessary to write a further book on using remote sensing imagery? We were both trained as ecologists, and our research has focused on resource conservation questions. As part of our work we have discovered that remote sensing image analysis offers a tool for working on regional-scale questions almost impossible to answer by any other means.

Remote sensing is any process that measures a phenomenon without actually coming into contact with it. In this book remote sensing imagery means any physical or computer-based representation of the radiation reflected from or emitted by terrain features or phenomena. Radiation detected by remote sensing systems is most commonly: (1) reflected sunlight, (2) heat generated by all objects that are warmer than absolute zero, or (3) reflected microwaves (radar). Images are produced by sensors that are placed on platforms such as cranes or booms that raise the sensor a few meters from the ground, or aircraft, or satellites that position the sensor at altitudes of tens to thousands of kilometers. Sensors (e.g., a 35mm camera or videocamera) are devices that create images by coupling optical systems that focus radiation from the terrain below, with detectors that chemically or electronically react to the intensity of the radiation, and with a system for recording the information generated by the detectors. Remote sensing systems can generate a visual representation of terrain features or phenomena within areas that, depending on the altitude of the platform, range from a few meters wide to a complete hemisphere of the globe. This extraordinarily flexible spatial coverage, coupled with almost instantaneous data collection that does not alter the feature or phenomenon of interest, makes remote sensing imagery an ideal tool for monitoring natural resources.

We started using remote sensing data not as engineers designing the technical specifications for the next generation of sensor systems but as applied biologists seeking answers to resource conservation questions. The books that guided us through our first tentative forays into remote sensing image analysis were of two kinds: (1) beautifully presented large-format books, replete with color plates that illustrate well the range of phenomena that can be remotely sensed but with text that provides little guidance for conducting an analysis, and (2) the comprehensive technical texts that describe in great detail the physics and mathematics behind remote sensing image analysis, but tend not to dwell on the limitations of the technology or the pitfalls to avoid.

Though we learned much from these texts and from the rich literature available in scientific journals, we still made many silly mistakes. We realized after talking with our colleagues, who had starting using remote sensing data under the same circumstances, that we had all made many of the same mistakes—mistakes that we passed off as part of the necessary "learning experience," but which we now realize should have been avoidable. We also realized that many of our mistakes were born of unrealistic expectations, caused in part by remote sensing information being "oversold" in many texts as an instant answer to all questions. In some cases, researchers had even stopped using this valuable tool because their simple errors and erroneous assumptions produced results that failed to meet their exceedingly high expectations.

In retrospect, we noticed that the methods sections of academic journal articles describe only the techniques that prove successful—not the approaches that fail. Negative results and failed methods are rarely accepted for publication, and thus first-time users of remote sensing imagery are likely to take the same wrong turns that experienced users have learned, the hard way, to avoid.

We wrote this book to fill a gap between the glossy application texts and the highly mathematical volumes. It is intended as an introduction to the practical use of remote sensing imagery to help solve problems in conservation biology and environmental management. We hope this book will give first-time users more realistic expectations about what remote sensing image analysis can and, most important, cannot do. We have tried throughout to provide practical advice and to emphasize key concepts, methods, and strategies essential for providing reliable answers to resource monitoring questions using remote sensing image analysis.

Our book is not the last remote sensing volume you should read, but we hope that it may be the first. Ideally, you will consider our advice (both cautionary and encouraging) before you begin your first serious study with remote sensing imagery—or, at least, before you allow a prior unhappy experience with image analysis to turn you away from this tool forever. We have learned a great deal from the writings of other scientists, and our book borrows heavily from their wealth of experience. The texts that have proven most valuable and informative to us are shown in table 1.1. We strongly suggest that you build a personal library from these texts if you intend remote sensing image analysis to become a frequent tool in your research.

Table 1.1
Key Reference Books in Remote Sensing Image Analysis

Cracknell, A. P. and L. W. Hayes. 1990. *Introduction to Remote Sensing.* London: Taylor and Francis.

Curran, P. J. 1986. *Principles of Remote Sensing.* London: Longman.

Harrison, B. A. and D. L. Judd. 1989. *Introduction to Remotely Sensed Data.* Canberra, Australia: Commonwealth Scientific and Industrial Research Organisation.

Hobbs, R. J. and H. A. Mooney. 1990. *Remote Sensing of Biosphere Functioning.* New York: Springer-Verlag.

Jensen, J. R. 1986. *Introductory Digital Image Processing.* Englewood Cliffs: Prentice-Hall.

Lillesand, T. M. and R. W. Keifer. 1994. *Remote Sensing and Image Interpretation.* 3rd edition. New York: Wiley.

Mather, P. M. 1987. *Computer Processing of Remotely Sensed Images: An Introduction.* New York: Wiley.

Richards, J. A. 1986. *Remote Sensing Digital Image Analysis: An Introduction.* New York: Springer-Verlag.

Sabins, F. F. 1986. *Remote Sensing.* 2nd edition. San Francisco: Freeman.

The Scale of Human Land Transformation

Smoke and anesthetized bees swirl around an Efe pygmy man, as he chops through the limb of a towering *Brachystegia laurenti* that hides one of the forest's treasures—honey. With an explosive crack the limb tilts earthward and crashes thirty meters through vines, shade-stunted saplings, and shrubs to burst apart on the ground below. Between May and August of each year, forager groups in this part of tropical Africa fell limbs and whole trees in search of honey. Yet from our viewpoint looking out the window of a Cessna as it flies a few hundred meters above the forest, it is impossible to tell if gaps in the canopy are natural tree falls or the result of honey gatherers.

For all but a fraction of the time since humans evolved, our resource consumption has left little visible impact on the landscape. Cutting trees for honey, burning grasslands for hunting, and gathering fruits and tubers were all at a scale that made them undifferentiable from natural disturbances such as tree falls, wildfires, and foraging by nonhuman fauna. Though it is true that foragers with relatively simple technology were capable of extirpating easily hunted species, low population density and seminomadic behavior meant that their observable impact on the landscape and its biological resources was slight. Humanity's relative invisibility on the landscape changed dramatically in neolithic times, perhaps 10,000 years ago. The advent of agriculture brought with it a more sedentary lifestyle, larger populations, and the capacity to alter, in clearly visible ways, the composition and spatial patterning of the landscape.

As the Cessna continues its course, the green cauliflower-like canopy of the forest occasionally breaks open into large, irregular clearings of barely vegetated soil or patches of swept sand speckled with huts. These openings are the fields and villages of Ituri forest farmers. The area of rain forest that we are flying over in northeastern Zaire has been occupied by farmers for more than a thousand years, and it is peppered by villages and clearings—some still in use, others abandoned recently and thus still visible.

Soon, the forest-village mosaic gives way to hundreds of hectares of oil palm plantation as we begin our brief descent to the airstrip. The drivers and passengers of four grumiers (logging trucks), stuck in the mud on the road below us, smile and wave as the plane skims over them. The pilot yells that they must be from the concession clearcut that we passed some sixty kilometers back. The landing warning squeals, and we jolt down the grass strip toward the immigration shack.

Our three-hour flight over the forest has mimicked, in a sense, the historical sequence of human impact on the content and coverage of ecosystems. The human capacity to modify the natural landscape has increased through time, accelerating with the exponential growth of human populations and the resource exploitation innovations associated with the agrarian and industrial revolutions. The plantations, logging concessions, and mining operations that we passed on our flight are highly visible evidence that even this isolated region of the world is not exempt from the growing global demand for raw materials. Indeed, our species is now the single most potent force shaping the world's biological and physical systems.

Today there are more than five billion of us. We consume directly (by eating, feeding livestock, clearing forest for lumber and fuelwood) and indirectly (through biomass burning to prepare agricultural fields, crop wastage, landscape conversion) as much as 40% of global terrestrial net primary production (Vitousek et al. 1986). The late twentieth century is manifesting a dramatic change in the relationship between people and the biological resources upon which we depend. With our ranks swelling by ninety million each year (Brown, Flavin, and Kane 1992), natural landscapes are being converted ever more rapidly. As a result, species are becoming extinct faster than at any other time, and climate appears to be changing. Extinctions and climate change, in turn, threaten both the function and productivity of the ecosystems upon which our health and welfare depend.

The changes in the rate and extent of resource exploitation that are happening in Zaire are a microcosm of the changes occurring everywhere. Not surprisingly, therefore, threats to biodiversity and ecosystem function and productivity are largely the same in the Ituri forest as in Antarctica or the Sundarbans mangrove forests of Bangladesh, the agricultural lands of central California, and New York's Central Park. Edward O. Wilson (1992) divided the threats to biodiversity into five major categories:

- habitat loss and fragmentation,
- habitat damage (pollution, disease, catastrophic disturbance),
- invasions by non-native species,
- overexploitation, and

- secondary extinctions resulting from the loss of critical (keystone) species.

To a great extent, all five categories of threats to biological diversity are a direct or indirect consequence of human use and abuse of earth resources. Human subsistence and market activities are bringing increasing distress to ecosystems around the globe, jeopardizing the structure, function, production, and resilience of earth's life-support systems. If these systems are not to suffer irreversible degradation, policy makers in government and the private sector must, as a first step, have available to them information sufficient to make sound decisions for sustainable resource management. Research into the rate and extent of ecosystem change, and the responses of ecosystems to stresses, is essential if strategies for ecosystem conservation and recovery are to be proposed, tested, and refined. The speed and scale with which humans are modifying earth systems limits the ability of traditional and intensive methods of field study to provide timely information over large areas. Ecologists, conservation biologists, and resource managers must look therefore to new tools, such as remote sensing image analysis, for gathering additional information relevant to ecosystem management and biodiversity conservation (Western and Pearl 1989).

Remote sensing provides a nonintrusive, multispectral perspective for viewing and measuring biological and physical phenomena, at multiple spatial and temporal scales. Remote sensing systems offer a unique and highly flexible tool to survey and monitor biophysical resources, to characterize the flow of energy and matter within ecosystems, and to track changes in the composition, extent, and distribution of communities and ecosystems (Quattrochi and Pelletier 1990). These three factors are at the root of the major ecological research questions considered by the Ecological Society of America to be critical to sound management of biological resources (Lubchenco et al. 1991). These questions pertain to

- biological diversity and ecosystem function (stability and resilience),
- impact of fragmentation,
- importance of corridors in a fragmented landscape,
- evaluation of stress,
- consequences of land- and water-use changes,
- patch persistence related to size, shape, and isolation,

- habitat classification and suitability,
- ecotone and boundary effects, and
- impact of juxtaposition of wildlands with man-altered systems.

Collection and analyses of remote sensing data, therefore, not only facilitates the synoptic analyses of earth-system function, patterning, and change at local, regional, and global scales over time; such data also provide a vital link between intensive, localized ecological research and the regional, national, and international conservation and management of biological diversity.

Summary of Chapters

Chapter 2 describes the basic physical principles upon which remote sensing of earth resources is based. We do not intend to bog down the reader in technical details. Rather, we aim to provide a basic understanding of the interaction of electromagnetic radiation and matter—an understanding essential for successful application of remote sensing data to problems in conservation biology. We stress the importance of knowing how sunlight and microwaves interact with the atmosphere and terrain features such as soils, water, and vegetation. We describe what sorts of phenomena can be remotely sensed, and, most important, what factors limit the ability of current technologies to remotely sense terrestrial and aquatic natural resources.

Chapter 3 discusses why remote sensing image analysis can be a powerful tool for conservation ecologists and natural resource managers. We describe the advantages and limitations of remote sensing imagery and discuss the trade-offs associated with aerial photography, satellite multispectral imagery, aerial videography, and radar. Last, we list the approximate comparative costs of remote sensing data that differ in spatial, spectral, radiometric, and temporal resolution.

Chapter 4 expands on chapter 3 by providing details of the most important and commonly used sources of remote sensing information. We describe the types of imagery available, how the imagery is generated, and how the information contained within the imagery differs from one sensor system to another.

Chapter 5 is the first of three chapters that in combination provide the first-time user with a guide to undertaking a study using remote sensing imagery. We begin this chapter by discussing the process of defining a problem that remote sensing analysis is a good candidate to help solve. We then describe how to select and acquire the most appropriate imagery. Finally, we demonstrate the most common tools and methods for displaying the images and correcting errors in the data.

Chapter 6 builds on the previous chapter by discussing how to increase the useful information contained within the imagery. We describe a variety of image enhancement techniques, such as contrast stretching, spatial filtering, and generation of vegetation indices, that are designed to emphasize features within the imagery of interest to the researcher and to improve one's ability to interpret what is seen.

Chapter 7 explains how to transform remote sensing images into thematic maps. We discuss the methods and the process of combining computer-based image analysis and visual interpretation of images with cost-effective field surveys, to generate land-cover or land use maps from remote sensing imagery. In this chapter we emphasize that the final maps created by a remote sensing analysis are only as good as the researcher's knowledge of the area of interest, and that field surveys are a key component of all studies that incorporate remote sensing imagery.

To show how imagery changes during processing, we have decided to use, as far as possible, one section of a Landsat TM scene of central Maine. In this way the first-time user can see how each manipulation of the image affects the information available and can easily contrast copies of the same image at different stages of the analysis. We believe that the best way to learn how to use remote sensing imagery is to go through the process of analyzing an image. To facilitate this we have made available copies of the imagery used in chapters 5, 6, and 7 and the software to view the analyses described throughout the chapters.

Raw and processed images used in the book and internet pointers to IBM PC and Macintosh image display software in the public domain are available via anonymous FTP at bandersnatch.fnr.umass.edu in directory pub/rs, or via the World Wide Web (WWW) at http://bandersnatch.fnr.umass.edu/pub/rs.html.

Chapter 8 exposes the reader to examples of how other researchers or agencies have made use of remote sensing information to enhance their knowledge of natural systems, to gain reliable estimates of anthropogenic or natural phenomena occurring at spatial or temporal scales that preclude on-the-ground investigations, and to improve resource management decisions. We constructed these case studies, in part, from articles published by the researchers, but we also interviewed the authors. We asked them to tell us about the problems they encountered during their studies and what they might have done differently in retrospect. They all complied graciously. Thus we are able in this final chapter to pass on their tips and clues to successful image analysis—words of wisdom that do not appear in their matter-of-fact journal articles.

2

What Is Remote Sensing and What Can It Reveal?

Interaction of Electromagnetic Radiation and Matter
Spectral Responses of Water, Soils, and Vegetation
Thermal and Other Factors Affecting Spectral Responses
Microwave Sensing of Terrain Features
What Sorts of Information Can Be Remotely Sensed?
Challenges in Using Remote Sensing for Resource Management

Until very recently, almost all of what was known about the form and function of earth's natural resources came from highly localized in situ studies. Scientists interested in a particular natural phenomenon typically selected and then visited a geographic area in which to study it, relying on direct observation or contact sensors to probe, for example, temperature, humidity, and soil pH. Remote sensing, in contrast, allows a scientist to gather information from a distance without coming into direct contact with the object or phenomenon of interest. By varying the height that the remote sensor is positioned above the earth, information can be obtained over a wide range of spatial scales. Moreover, remote sensing systems, such as aerial photography, satellite imagery, and radar, were not designed only to answer single, specific questions. Rather, the information that they gather about features on or near the earth's surface has many applications. A remote sensing image obtained for, let's say, Yellowstone National Park in Wyoming, contains information about the landscape that can assist foresters to identify areas of high fire risk, park rangers to quantify tourist impacts, geologists to map soils and rock formations, archaeologists to locate old road systems, and biologists to estimate grass biomass available for elk grazing. The general purpose nature of most remote sensing systems allows the information to be shared by many researchers from different disciplines, providing considerable bang for the buck.

Most technologies used for remote sensing obtain information about natural phenomena by measuring electromagnetic radiation (EMR). The EMR of interest may be natural, such as reflected solar radiation or the EMR emitted by warm-bodied animals, forest fires, and sun-warmed rocks and soils. Or it may be artificial, such as the reflectance from radar. Because EMR travels at the speed of light, remote sensors, unlike direct field studies, are able to record information about phenomena almost instantaneously over a wide range of spatial scales (table 2.1).

The term *remote sensing* was not coined until the early 1960s. Nevertheless, remote sensing information has been used to describe visible EMR emanating from landscapes ever since humans began recording the colors and shapes they saw in drawings and paintings (table 2.2). Automated recording of reflected EMR became possible in 1839 with Louis Daguerre's invention of light-sensitive copper plates. Subsequent development of color photography and of false-color camouflage detection film for infrared (IR) wavelengths expanded the range of remote sensing applications for military and natural resource management uses. Nonphotographic sensors (photoconductors, photodiodes, and charge-coupled devices) expanded the EMR range that could be detected, and allowed sampling within more narrowly defined wavebands (Norwood and Lansing 1983). Kites, carrier pigeons, balloons, airplanes, and space vehicles freed remote sensing devices from the confines of imaging from ground level, increasing greatly our regional and hemispheric view of the world and its components. Plate 1 presents the electromagnetic spectrum and the sensitivity range of different kinds of remote sensing systems.

Interaction of Electromagnetic Radiation and Matter

The sun is the primary source of EMR that strikes the earth's atmosphere and terrain features. Almost 50% of the radiation energy generated by the sun falls within the visible wavelengths of the spectrum (0.4–0.7μm), with a maximum in the blue region at 0.47μm (see plate 2). When EMR strikes an object, the energy may be:

1. **Transmitted** - energy passes through the object but its velocity is changed (refracted). As the frequency of EMR radiation remains constant once emitted from its source, a change of velocity as a result of refraction induces a change in EMR wavelength according to the equa-

Table 2.1

Examples of Remote Sensing Systems

Sensor System	Platform	Imagery Type	Spectral Resolution	Spatial Resolution	Temporal Resolution	Spatial Coverage
Meteosat	Geosynchronous orbit	Digital - pixel mosaic	Visible to thermal IR	2.4km	30 minutes	hemispheric
NOAA AVHRR	Polar orbiting	Digital - pixel mosaic	Visible to thermal IR	1.1 to 4km	12 hours	2700km
Landsat MSS	Polar orbiting	Digital - pixel mosaic	Visible to near IR	79m	16 to 18 days	185 km
Landsat TM	Polar orbiting	Digital - pixel mosaic	Visible to thermal IR	30m	16 days	185 km
SPOT Panchromatic	Polar orbiting	Digital - pixel mosaic	Visible	10m	5 to 26 days	60km
SPOT HRV	Polar orbiting	Digital - pixel mosaic	Visible to near IR	20m	5 to 26 days	60km
ERS-1	Polar orbiting	Digital - radar	n/a	30m	16 to 18 days	100km
Radarsat	Polar orbiting	Digital - radar	n/a	10 to 100m	3 to 24 days	45 to 500km
Large format camera	Orbiting space shuttle	Analog photography	Visible to near IR	5 to 10m	during shuttle flights	
Photography	Booms and Aircraft	Analog photography	Visible to near IR	>0.25m	archive or on demand	<20km
Videography	Booms and Aircraft	Digital - frame	Visible to near IR	>0.25m	archive or on demand	<5km
Digital photography	Booms and Aircraft	Digital - pixel mosaic	Visible to near IR	>0.50m	archive or on demand	<20km

Table 2.2
EMR Visible Spectrum

Color	Waveband (μm)
Violet	0.40–0.43
Indigo	0.43–0.45
Blue	0.45–0.50
Green	0.50–0.57
Yellow	0.57–0.59
Orange	0.59–0.61
Red	0.61–0.70

tion $c = f\lambda$, where c is the speed of light, f the frequency, and λ the wavelength of the EMR.

2. **Absorbed** - energy is transferred to the object, usually in the form of heat.
3. **Reflected** - energy rebounds unchanged from the object, with the angle of reflection equal to the angle of incidence. Reflectance of an object is the ratio of reflected to incident radiation. Wavelengths reflected from an object (i.e., not absorbed) determine its color.
4. **Scattered** - the travel direction of energy is changed randomly. The degree of scattering is related to the wavelength of the EMR and the size of the objects that it strikes.
5. **Reradiated** - energy is first absorbed then re-emitted, most often as thermal (heat) radiation (see figure 2.1).

The EMR emitted by the sun covers a broad range of wavelengths—from high frequency, short wavelength gamma rays through X rays, ultraviolet, visible light, infrared, and microwaves, and finally to low frequency, long wavelength radio waves. However, not all solar radiation wavelengths reach the earth's surface, and therefore not all wavelengths can be used to remotely sense features or phenomena on the earth's surface (figure 2.2 and table 2.3). Water and several gases, primarily ozone (O_3), oxygen (O_2), and carbon dioxide, interact with and effectively block distinct wavelengths of radiation. Consequently, solar energy available to interact with surface features is divided into discrete windows, separated by "blank" regions. These windows appear as peaks in atmospheric transmission depicted in figure 2.3.

Each form of matter on earth transmits, absorbs, reflects, scatters, or

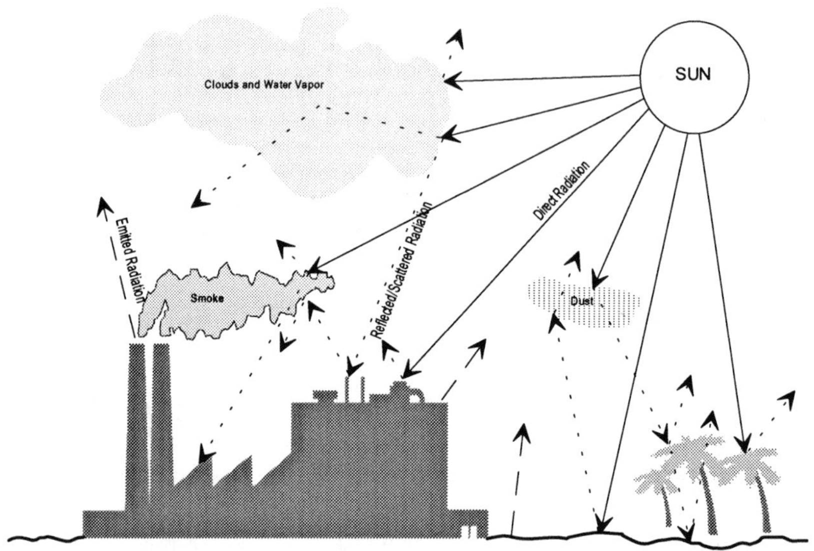

Figure 2.1 Fate of solar EMR as it interacts with the atmosphere and terrain features. Notice how terrain features and moisture, dust, and smoke in the atmosphere not only reflect radiation in different ways but also change its composition.

reradiates solar radiation in a particular and characteristic manner (figure 2.4). Black objects, for example, absorb all visible wavelengths from 0.4μm (violet) to 0.7μm (red). Red objects absorb short wavelengths (blues and greens) and reflect the longer red wavelengths. Objects that absorb primarily green wavelengths, reflecting light in both the blue and red parts of the spectrum, appear to us as purple. The radiation detected from an object (if we ignore atmospheric and sensor effects) is determined by its pigmentation, water content, texture, and whether it is situated in direct sunlight or shadow.

To interpret remote sensing images that constitute an almost instantaneous snapshot of the EMR reflected or emitted from an area of terrain, it is absolutely essential to begin with a basic understanding of how typical terrain features such as water, soils, rock and vegetation interact with EMR at different wavelengths. We cannot stress enough the importance of understanding how terrain features reflect, transmit, absorb, and emit EMR. For without knowing how the landscape interacts with EMR, and

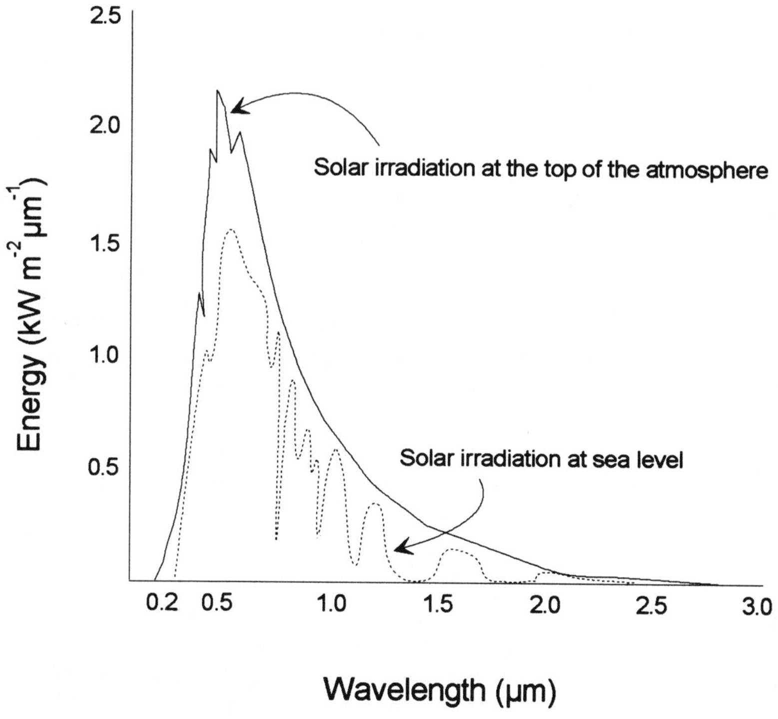

Figure 2.2 Solar irradiation and atmospheric absorption by ozone, oxygen, carbon dioxide, and water vapor.

how the atmosphere, sun angle, and topography alters EMR reflected, transmitted, and emitted from the terrain, it is almost impossible to interpret successfully the information obtained by remote sensing systems. For example, through experience we all know that water often appears blue, soils red or brown, and living vegetation green. Yet we must understand *why* water looks blue to the human eye, if we are to understand what is likely to happen to the color of a water body that is sediment or algal laden, polluted, or merely shallow.

Spectral Responses of Water, Soils, and Vegetation

Water Clear water appears blue to the human eye because it reflects EMR energy primarily in the short, blue wavelengths (0.4–0.5 µm) of the

Table 2.3
EMR Bands and Atmospheric Absorption

Waveband	Wavelength	Comments
Gamma ray	< 0.003nm	Radiation from the sun, completely absorbed by the upper atmosphere.
X-ray	0.03–3nm	Completely absorbed by atmosphere, thus not available for remote sensing.
Ultraviolet (uv)	3nm–0.4μm	Transmitted at wavelengths greater than 0.3μm, but atmospheric scattering is severe; detectable with film and electronic detectors.
Visible	0.4–0.7μm	Detectable with film and electronic detectors; earth reflectance peak at 0.5μm.
Infrared (ir)	0.7–300μm	Absorption varies with wavelength; peak transmission windows occur at 2–2.5μm, 3.5–4.5μm, 8–9.5μm, and 10–12.5μm.
—reflected ir	0.7–3μm	Good transmission in this range except for H_2O absorption at 0.9μm; ir film is sensitive in the 0.7–0.9μm range.
—thermal ir	3–15μm	Emitted radiation recorded using electronic detectors and not film.
Microwave	0.3–300cm	Wavelengths in this range are not absorbed by the atmosphere.
—radar	0.8–100cm	Radar is an active form of microwave remote sensing.

NOTE: 1μm (micrometer) = 1000nm (nanometers) = 10^{-6}m

visible spectrum. In contrast, clear water will appear completely black in the reflected IR region because it absorbs all radiation between 0.8–3.0μm. If water is carrying suspended solids or is shallow, its reflectance will change according to the quantity and nature (soils, algae, etc.) of sediments, and the composition of the bottom substrate (mud, sand, rocks, vegetation, etc.).

Figure 2.4 shows that even though clear and turbid waters share the same shape of reflectance curve, turbid water reflects more EMR at all visible wavelengths. For studies of water quality, depth, and turbidity we need to employ sensors that can gather reflectance data in the blue and green region of the visible spectrum (0.4–0.6μm). In contrast, we can use the fact that water appears black in the reflected IR region to detect the

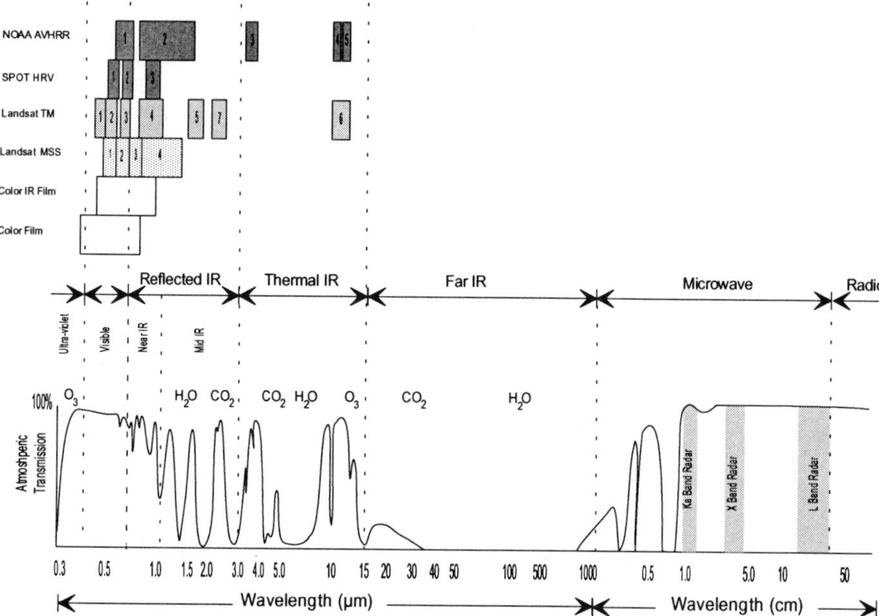

Figure 2.3 Atmospheric transmission windows and the wavebands of common remote sensing systems. Ozone, water, and carbon dioxide account for most of the blockages in transmission, as shown. The numbered boxes for each of the six remote sensing systems are the designated names for the wavebands sensed. Notice how the wavebands sensed by each system align very closely with areas of maximal atmospheric transmission (atmospheric windows).

interface between water and the land surface, especially if the land reflects radiation strongly in this range (0.8–3.0μm).

A major limitation of using remote sensing data for the study of water-related phenomena is that atmospheric interference (clouds, fog, and haze) is greatest in the blue region of the spectrum (0.4–0.5μm) and only a little less severe in the green region (0.5–0.6μm). Thus water in the atmosphere obscures our view of water-related (and other) features on the earth's surface.

Soils and rock Soils constitute the primary background spectral response in most vegetated landscapes, and, in combination with rock, the sole spectral response in nonvegetated terrain. It is important therefore to

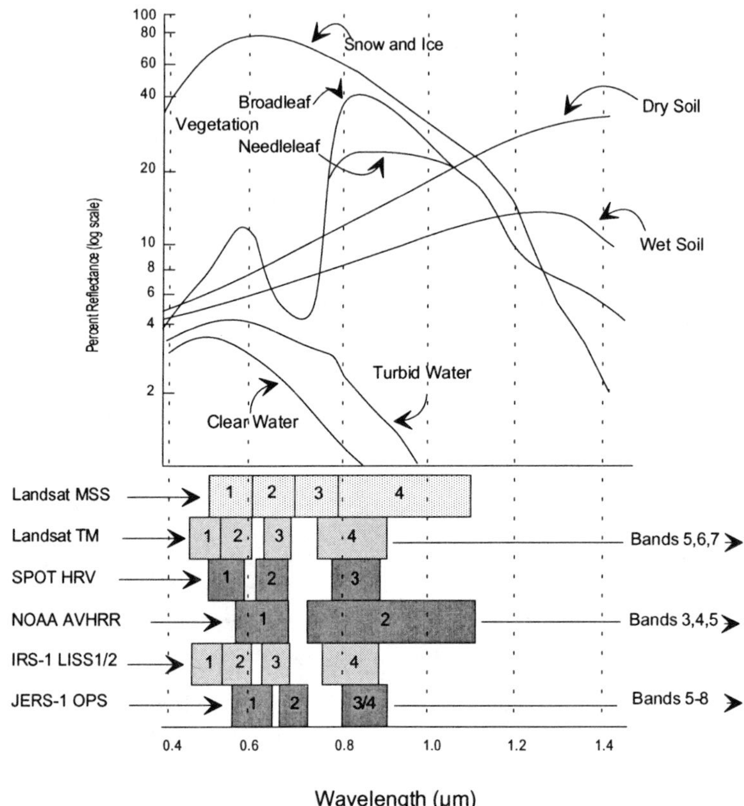

Figure 2.4 Generic spectral reflectance characteristics of soil, water, and vegetation, relative to the visible and near-IR wavebands of common remote sensing systems. Note how all sensor systems employ detectors that match the wavebands within which vegetation exhibits the strongest response (e.g., Landsat TM bands 2,4; JERS bands 1,3/4).

understand the impact of soil pigmentation and moisture content on spectral response. Dry, tan, silty soils tend to show a progressive increase in reflectance at longer wavelengths. In contrast to water, soils exhibit a bright response in the visible red (0.6–0.7μm) and reflected IR (0.7–1.1μm) regions of the spectrum (figure 2.4). As the moisture content of soils increases, in for example seasonally flooded fields, spectral reflectance changes progressively from a characteristic soil response toward a characteristic water response. As soil reflectance tends to increase monotonically from the blue region of the spectrum to longer wavelengths, it should be possible to detect soils by looking for features whose spectral reflectance increases consistently with wavelength.

Vegetation To our eyes, lush vegetation is green because chlorophyll, the primary pigment that drives photosynthesis in plants, absorbs solar radiation primarily in the blue wavelengths (peaks at 0.41, 0.43, and 0.453μm) and red wavelengths (peaks at 0.642 and 0.662μm), allowing green light to be transmitted through, and reflected from, the leaves. That the primary chlorophyll absorption peaks are located in a particular region of the blue part of the spectrum should not be surprising, given that 4.7μm carries the sun's maximum energy output. Interestingly, the radiation absorption properties of chlorophyll are so unique, and greenness is so characteristic of vegetation, that we all tend to consider most unidentified green objects, at least initially, to be plants or derived from plants. The twin spectral zones of chlorophyll absorption yield an overall vegetation spectral response that, in the short waveband, is very similar to that of soils—low in the blue region, with a gradual increase in reflectance through the green region. However, in the red absorption region (0.58–0.68μm), vegetation reflectance declines considerably, whereas soil reflectance continues to increase (figure 2.4). Furthermore, leaf cell structure scatters and reflects nonvisible, near-infrared radiation, providing an effective defense against overheating. As a result, radiation reflectance of vegetation increases dramatically from the chlorophyll absorption peaks in the red visible region, reaching a maximum in the near-IR between 0.8 and 1.1μm.

Because plants absorb light strongly in the blue and red visible regions of the spectrum and because they reflect strongly in the near-infrared, we can use this characteristic spectral response (signature) to distinguish plants from nonliving features. Even more important, we can use changes in reflectance to assess the composition, maturity, and health of natural vegetation and crops. This is because pigment (chlorophyll, carotenes, etc.) and water content vary according to species, according to growth stage (young and verdant, or dying and yellow), and when plants are diseased or under stress from drought.

Thermal and Other Factors Affecting Spectral Responses

In addition to reflecting radiation, all objects warmer than absolute zero (−273.16°C) emit radiation in the thermal infrared region of the EMR spectrum. The actual surface temperature of an object can be determined remotely, using the Stefan-Boltzmann law, by measuring the thermal infrared radiation that it emits. This method works, however, only if the object approximates a "perfect blackbody emitter" with uniform reflectance. Only water satisfies this requirement. Thermal remote sensing can thus be used to determine the temperature of water bodies.

Because of remote sensing, it is now possible to regularly measure sea surface temperatures over large areas. Such information is vital to climate modelers and meteorologists; it can indicate the location and direction of ocean currents and warn of shifts in marine and coastal primary production. Thermal emissivity can also be used qualitatively to discriminate features within terrestrial landscapes. For example, Sidle et al. used an airborne thermal IR sensor to survey sandhill cranes roosting along the Platte river in Nebraska (Sidle et al. 1993), and Barber et al. (1991) used a similar technique to count walrus in Alaska.

Knowledge of the wavelengths at which atmospheric features absorb, transmit, or reflect EMR, combined with an understanding of the distinctive spectral characteristics of water, vegetation, soils, and rocks, enables engineers to design remote sensing systems that can gather earth resources information valuable to, inter alia, ecologists, geographers, foresters, agriculturalists, marine biologists, geologists, and planners. (Table 2.4, for example, lists the general kinds of terrain features that can be distinguished by the seven bands of Landsat TM imagery.) Similarly, researchers interested in using remote sensing imagery can use their knowledge of the spectral response of terrain features to select the most appropriate satellite sensor and wavebands to discriminate important features from one another, and from the background.

Though the spectral response of a terrain feature may in many cases be characteristic (and hence diagnostic) of that feature, the response may be mutable. The spectral responses of all terrain features tend to be affected by a broad range of biophysical and sensor-related factors (table 2.5). It is exceedingly important therefore to understand how changes in sun angle, plant growth stage, soil moisture, etc. are likely to alter the spectral response of terrain features of interest. For example, suppose you are interested in mapping all active slash-and-burn cultivation areas within a particular moist tropical forest. The spectral response of active fields will vary greatly depending on the stage in the cultivation cycle (felled, burned, cleared, sown, crop covered, harvested, and abandoned). With each phase in the agricultural cycle the proportion of soil to vegetation visible to the satellite sensor changes, as does the dominant crop and

Table 2.4
Attributes of Landsat TM Spectral Bands

Band 1 (0.45–0.52μm; blue)—Lower range encompasses peak transmittance of clear water and is thus useful for water quality studies and for bathymetric studies in which maximum light penetration is desired. Upper range bounds peak chlorophyll absorption, thus making it useful for distinguishing soil from vegetation, and deciduous from coniferous plants.

Band 2 (0.52–0.60μm; green)—Captures green reflectance peak of healthy vegetation; useful for assessing plant vigor.

Band 3 (0.63–0.69μm; red)—Corresponds to upper chlorophyll absorption peak, making it one of the best bands for discriminating vegetation types. This band exhibits more contrast than do bands 1 and 2, as it penetrates atmospheric haze more effectively.

Band 4 (0.76–0.90μm; reflected ir)—Corresponds to reflectance maximum of vegetation, making it useful for plant biomass studies, as well as soil-crop and land-water boundary discrimination.

Band 5 (1.55–1.75μm; reflected ir)—Indicates moisture content of vegetation and soils, making it useful for discriminating vegetation types and for plant vigor studies. It is also used to discriminate among clouds, snow, and ice.

Band 7 (2.08–2.35μm; reflected ir)—Corresponds with absorption wavelengths for hydroxyl ions in minerals. The ratio of bands 5 and 7 is useful for mapping hydrothermically altered rocks associated with mineral deposits.

Band 6 (10.4–12.5μm; thermal ir)—Measures infrared radiation emitted from the terrain, thus making it useful for locating geothermal activity and for assessing vegetation stress, biomass burning, and soil moisture. Even though this band records wavelengths longer than those recorded by band 7, it is called band 6 because band 7 was designed into the sensor system after band 6.

SOURCES: Sabins 1986:86; Jensen 1986:34.

the color and size of leaves. Furthermore, as the seasons change, so too will (a) sun angle—changing illumination and shadow effects; (b) cloud and haze patterns—obscuring or attenuating light reflected from the terrain; and (c) rainfall and thus soil and vegetation moisture content—altering IR reflectance characteristics of the terrain.

For example, if we obtain remote sensing imagery when the fields are heavily vegetated with crops, we may be unable to distinguish the spectral response of these fields from the background regrowth forest. However, if we obtain imagery when the fields are burned and bare, the field's spectral response will be dominated by soil and ash and thus is much more likely to be very different from the surrounding vegetation. Similarly, to separate hardwoods from softwoods in New England we can take advantage of

Table 2.5
Sources of Variation in Radiation Detected from Earth

Illumination Conditions	**Site Background Conditions**
Sun angle and cloud distribution	Weather
Spectral range of source radiation	Hydrology
	Soils
Atmospheric Conditions	Topography
Water vapor, aerosols, dust, etc.	Surrounding features
Plant EMR Interactions	**Observation Conditions**
Plant architecture, density, and distribution	View angle relative to radiation source
Pigmentation	Time of day
Thermal emittance	Altitude
Plant Condition	**Sensor Effects**
Maturity	Electronic noise, gain change
Variety	Calibration precision
Turgidity	Intersystem variability
Health/Disease	Instantaneous field of view

SOURCE: Adapted from Barrett and Curtis 1982:44.

the distinctive autumn leaf coloration of hardwoods, or the fact that they drop their leaves in the winter.

If the area of interest has rugged topography, it will be important to select a source of remote sensing imagery with a high sun angle—reducing shadows. The Landsat satellites orbit the earth from pole to pole. Imagery is obtained only during the north to south portion of the orbit when the satellite is in the sunlit side of the earth. The return segment of the orbit from south to north is in the dark, away from the sun. Imagery from any of the Landsat satellites designated 1–5 is obtained during the early hours of the day (the satellites fly over the equator at 9:30 A.M.). It thus suffers from more severe topographic shadow effects than does imagery from the SPOT satellites that cross the equator at 10:30 A.M.. Clearly, the more we understand about what influences the spectral responses of terrain features, the more able we will be to make sage decisions as to the type and timing of remote sensing imagery to obtain, and the more accurate our interpretations will be.

Microwave Sensing of Terrain Features

With the recent and planned launch of several satellites equipped with microwave remote sensing systems, longer wavelength data are now becoming more readily available for natural resource management applications. Microwave terrain imaging systems, notably radar, have been used by the U.S. military since the 1950s. Though acquisition methods are well advanced, experience in analyzing these images for natural resources applications is considerably less well developed when compared to visible and IR image interpretation. Microwave sensors are likely, however, to be extremely useful for some natural resource management applications, as they offer a fundamentally different view of the world than that of visible and IR remote sensing. Microwave remote sensing is distinctive because:

- Microwaves are largely unaffected by atmospheric conditions, and are thus able to "see through" smoke, clouds, haze, and snow.
- Microwaves interact with terrain features very differently than do visible and thermal EMR.
- Microwave sensing is primarily an active process, in that the terrain is illuminated not by the sun but by energy supplied by the sensor.

The microwave region of the EMR spectrum includes wavelengths ranging from approximately 1 millimeter (mm) to 1 meter (m)—far longer than the micrometers (μm) of the visible spectrum and the nanometers (nm) of X rays and gamma rays. In the microwave region of the spectrum, naturally emitted or reflected microwave energy from terrain features occurs only at extremely low levels. Microwave remote sensing has thus focused mostly on active systems such as RADAR (RAdio Detection And Ranging), which was developed during the 1940s under the press of war. Active microwave systems map terrain features by transmitting a series of microwave pulses and recording the strength and timing of echoes reflected from objects in the system's field of view.

Microwaves are only slightly attenuated by atmospheric conditions such as clouds, smoke, and haze. Heavy rain does reflect wavelengths shorter than 3cm—a characteristic used by meteorologists to map rainfall in real time using the radial sweep radar familiar to viewers of nightly news broadcasts. The strength of the signal that returns from an object is a function of sensor parameters, such as wavelength, polarization, and incidence angle, and of object parameters, such as roughness, size, shape and orientation, electrical reflectivity or conductivity, and water content (figure 2.5).

The geometry of terrain features affects what proportion of the transmitted signal is returned to the microwave sensor (antenna). *Diffuse reflectors,* such as vegetation, tend to reflect the signal in all directions, so

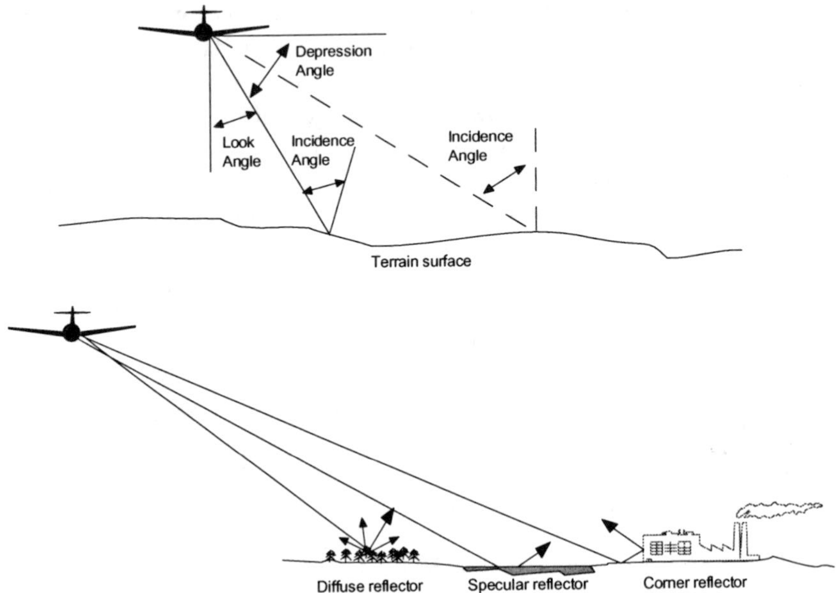

Figure 2.5 Microwave (radar) terminology and reflection from terrain surfaces.

only a small portion returns to the sensor. Even less is picked up by sensors when a signal strikes a smooth, mirrorlike surface because most of the transmitted energy is reflected away from the sensor. This kind of smooth terrain feature is called a *specular reflector,* and it produces a very low intensity return signal (backscatter), corresponding to a dark tone in the final image. The brightest portions of images produced by microwave sensing are terrain features that are metal objects or whose geometries act as *corner reflectors.* The juxtaposition of two smooth surfaces, as in the wall of a building and the smooth ground around it, causes a double, or corner, reflection (see figure 2.5) that returns a strong signal to the sensor.

A vegetation canopy interacts with microwaves as a group of discrete components (leaves, branches, trunks, etc.). Vegetation, like snow and rainfall, is considered a volume rather than a surface scatterer of microwaves. Microwave reflection from vegetation is greatest when the radar wavelengths are close to the average size of vegetation components. Thus, short 2–5cm microwaves are best for sensing crops and tree leaves. If the vegetation canopy is sparse or the radar wavelengths are longer (10–30cm), tree trunks and soil surface will contribute more microwave backscatter than will the leaf canopy. The signal strength reflected from vegetation can be strengthened by using cross-polarization—that is, by transmitting microwaves with horizontal (H) polarization and receiving the backscatter with vertical (V) polarization. In contrast, like-polarized radar (HH or VV) can penetrate vegetation to sense the terrain below. In vegetation-free areas, random roughness of the surface, periodic surface pattern (e.g., furrows), and soil texture will all affect the strength of microwave backscatter.

The electromagnetic characteristics of objects also affects the extent of microwave backscatter. An object's reflectivity is in part a function of its *complex dielectric constant,* that is, a measure of the electrical property of matter that influences radar returns. Objects with a high dielectric constant reflect a greater proportion of transmitted microwaves, appearing as brighter tones in the final image. Metals and water reflect microwaves strongly because they have large dielectric constants (>80). Whereas, the dielectric constant of dry vegetation and soils is only in the range of 1–10. As the water content of soils or vegetation increases, so does the reflectivity. In contrast, as pure water becomes brackish or sediment laden, its reflectivity declines.

What Sorts of Information Can Be Remotely Sensed?

Remote sensing imagery can provide basic measurements of a range of biological and physical characteristics of a landscape, such as position, shape, elevation, color, temperature, and moisture content (table 2.6). The most important forms of basic information of remotely sensed biophysical characteristics of a landscape are all measured on interval or ratio scales. For example, temperature is determined using the (interval) value of a single thermal sensor, whereas biomass often is assessed using the ratio of the values of the reflected IR and red sensors.

When the basic measurements are combined, hybrid (also called synthetic) information can be extracted from the imagery, thus describing the landscape in new and useful ways. The most often used synthetic information extracted from remote sensing imagery is land cover (or land use), which is determined by combining information on object color, shape,

Table 2.6
Basic Information Available from Remote Sensing Imagery

	Visible			IR		
Biophysical Feature	Blue	Green	Red	Reflected	Thermal	Radar
x,y position	✓	✓	✓	✓	✓	✓
shape, size, orientation	✓	✓	✓	✓	✓	✓
topographic elevation (stereo images)	✓	✓	✓	✓	✓	✓
bathymetric depth	✓					
subsurface location						✓
color	✓	✓	✓	✓		
temperature					✓	
texture/roughness	✓	✓	✓	✓	✓	✓
Vegetation						
chlorophyll absorption	✓		✓			
biomass				✓		
moisture content				✓		
Soil						
moisture content				✓	✓	✓
Minerals						
hydroxyl ion content				✓		
Fires					✓	

location, size, etc. with the user's knowledge of the landscape being examined. Other composite variables are shown in table 2.7. Extraction of useful, quantitative biophysical data and information on synthetic characteristics of terrestrial and aquatic landscapes demands an understanding of how electromagnetic radiation is absorbed, reflected, and emitted from objects in the landscape, and how the radiation reflected or emitted from landscape objects is altered and attenuated by the intervening atmosphere and by the spatial and spectral resolution of the remote sensors themselves.

Challenges in Using Remote Sensing for Resource Management

Natural resources managers are accustomed to working with information that is not perfect. It should come as no surprise, therefore, that information about the terrain that would be useful or of interest is often attenu-

Table 2.7
Synthetic Information Extracted from Remote Sensing Imagery

Renewable Resources	Nonrenewable Resources	Planning/Management
leaf area	topography	land use/cover
crop inventory	landforms	land use/cover change
growing season	lithology	environmental impact
range forage condition		human population density
soil classification		surface currents
soil erosion		coastal zone monitoring
soil moisture		bathymetry
forest inventory		water quality/pollution
vegetation health/condition		drainage patterns
wildlife habitat evaluation		flood estimation
		landscape diversity/patchiness

SOURCE: Adapted from Quattrochi and Pelletier 1990.

ated or obscured within remote sensing imagery as a result of the resolution limitations of the sensor, shadows from topography or low sun angle, and atmospheric interference such as clouds, haze, smoke, and dust. To remove noise from imagery or to minimize noise by selecting the most appropriate sensor and image acquisition time, it is vital that we understand the factors that combine to generate information or noise within a remote sensing image.

The amount of radiation from the terrain recorded by a sensor and stored as a remote sensing image is a complex function of

- the characteristics of the remote sensor (spectral, radiometric, spatial, and temporal resolution);
- the size, shape, color, orientation, chemical composition, and water content of terrain objects such plants, buildings, rivers;
- the density, distribution, and juxtaposition of terrain features within the landscape; and
- the noise introduced by the intervening atmosphere.

The more we know about how these factors affect the information available within the image, the better we will be at characterizing and analyzing remote sensing imagery. Monet, the great French impressionist painter, with his numerous studies of corn fields, ornamental gardens, and the church of Notre Dame, understood well how the same scene viewed at different times, in different seasons, under different lighting conditions could look very different. By conveying slightly or vastly different information each time the same scene was painted, Monet could evoke a range of feelings and interpretations.

Users of remote sensing information must be aware, as was Monet, that the same area of terrain can look very different at different scales, in different seasons, and under different lighting conditions. A typical section of western Massachusetts, with its patchwork of rolling wooded hills interspersed with croplands and pastures and river valleys, puts on a different face with each season. In the early spring: (1) the low sun angle typical of northern latitudes in winter will cast heavy shadows in hilly areas and along the edges between forested areas and croplands; (2) trees are without leaves, the annuals have yet to emerge, and the soil is saturated with recently melted snow; (3) the air is crisp and dry—with few clouds and low humidity. From the air, the pale bark of stands of birch and aspen are in stark contrast to the dark brown patches of oak and beech forest, and the greens of spruce and pine woodlands. As the seasons change from spring to summer to fall, sun angle, humidity and haze levels, leaf biomass and color, and soil moisture all change, as does the information available within remote sensing imagery.

It cannot be overstated that the more we know about how the atmosphere and terrain features interact with solar radiation, the more likely it is that we will be able to interpret accurately the information contained within remote sensing imagery. There is no substitute for conducting field surveys and getting to know your study area. Whenever possible, try to view the area captured by the remote sensing image from the air, as a vertical viewpoint is vastly different from that of a ground-level tangential perspective. Understanding how leaf area, leaf age, canopy closure, herbaceous vegetation, and soils combine to generate a composite terrain feature reflectance, and how that composite reflectance is likely to change with the seasons, will greatly improve one's capacity to interpret the radiation brightness information available within aerial photographs, aerial videography, and satellite imagery.

Knowing how vegetation cycles, terrain feature architecture, topography, weather patterns, sensor resolution, and time of day and year can affect the information content of remote sensing imagery is vital for: (1) selecting the most appropriate source of data and timing of data collection, (2) removing noise from remotely sensed images, and (3) characterizing and analyzing the information contained within images.

3

Remote Sensing for Monitoring and Managing Natural Resources

Problems in conservation biology almost always have a geographic component. We routinely need to know the distribution of vegetation, human land use, water resources, or soils within a given region of the globe. Knowledge about the distribution of features has traditionally come from intensive field investigations of relatively small areas, the results of which are then extrapolated to a regional scale. As the rate, magnitude, and complexity of human impact on natural systems continue to increase, we are going to have to find ways to gather information about the environment that can

- provide us with as much or more detail than field surveys,
- accomplish this over large areas, and
- do so repeatedly and at moderate cost.

Remote sensing data offer tools to extend the knowledge gained from intensive in situ ecological research to larger geographic areas, at more frequent intervals, and over a longer time series. Remote sensing takes advantage of the fact that all kinds of matter reflect or emit electromagnetic radiation, and we can thereby detect and measure biological and physical features at or near the surface of the oceans, on land, and throughout the atmosphere. Most features detected by remote sensing can be measured in other ways. However, remote sensing data can be obtained at a wide range of scales and at multiple times, providing a unique source of information on ecological phenomena synoptically over time.

The terms *large-scale* and *small-scale,* when applied to maps and remote sensing images may, at first glance, seem counterintuitive. For example, map makers and users of remote sensing imagery refer to maps (or images) of 1:250,000 as small-scale and maps of 1:1,000 as large-scale. This is because scale refers to the ratio of the size of a landscape feature as represented within the map or image to its true size. A map at 1:1,000 (that is, in which one unit of length on the map corresponds to a thousand units on the ground) has a map-to-ground ratio of 0.001; a map at 1:250,000 has a map-to-ground ratio of only 0.000004. The 1:1,000 map thus has the "larger" scale. More simply, scale determines the size that an object or feature will appear on the map or image. For example, a baseball diamond will appear very small (if at all) in a small-scale map, and much larger in a large-scale map. Increasing the scale of a map will increase the size of objects within the map, and the detail that is visible within the map.

Remote sensing allows local or global surveys to be conducted instantaneously and presented in maplike snapshots or images. Developed initially for secret military surveillance, remote sensing information and data analysis techniques have become progressively more accessible to the public during the past twenty years. Today conservation biologists and environmental managers have a remarkably powerful toolbox with which to map natural resources and to monitor the changes that ecosystems are undergoing.

Remote sensing is particularly useful for environmental studies because:

- Observation or detection of a phenomenon by means of remote sensing does not result in a change in that phenomenon (i.e., there is no observer interference).
- Data are gathered almost instantaneously, regardless of spatial scale, and data can be obtained frequently over long time periods.
- A wide variety of sensors with different spatial and spectral resolutions are available to gather information.

What are the major characteristics of remote sensing information that gives it such puissance for identifying and finding solutions to environmental problems?

Spatial, Temporal, and Spectral Advantages

Remote sensing data offer remarkable flexibility in the perspective from which we view the world. We can view the spatial patterning of individual wombat colonies in Australia (Loffler and Margules 1980), observe the seasonal greening and senescence of vegetation over the whole of continental Africa (Tucker et al. 1985), or generate a global map of nine basic vegetation types (Koomanoff 1989). This immense range in spatial coverage is one reason why remote sensing imagery is such a powerful descriptive and analytical tool.

Remote sensing information dramatically increases the scale and range of our world view. Aerial photography, videography, and satellite imagery can provide highly detailed information about a given area, or allow us to see huge areas of the world at one time (plate 3).

Remote sensing imagery obtained from aircraft or satellites allows us to map and revisit areas on a time scale varying from a matter of hours to decades. This is an exceedingly important characteristic of remote sensing data because detecting and monitoring change is one of the most commonly encountered challenges that face conservation ecologists. Change may be sudden and short-lived, requiring swift data acquisition and analyses to monitor and evaluate the impact of events such as storms, effluent discharge, locust plagues, and droughts. Remote sensing provides the capacity to obtain imagery rapidly and repeatedly over large areas, thus permitting a broad range of environmentally important but transient phenomena to be monitored.

The proportion of a region flooded by a recent storm, for example, may be very difficult to determine on the ground, but could be assessed quickly and accurately using aerial photography, videography, or satellite imagery (Blasco et al. 1992; Brown et al. 1987; Sidle et al. 1992; Barton and Bathols 1989). Similarly, the size, velocity, and dispersion rate of an oil slick (Cross 1992), the geographic distribution and timing of savanna burning in the Sahel (Prince et al. 1990), the development and movement of phytoplankton blooms in estuaries (Tyler and Stumpf 1989), or the acreage of hardwood forest in New England blown down or damaged

Table 3.1
The Spatial Range of Management Uses of Remote Sensing

Individual Item Massachusetts town planners, concerned about the impact of urban pollution on tree survivorship, are able to map the location and to measure the canopy area of all trees within the central business districts of a metropolitan area by using high resolution aerial videophotography (figure 3.1).

Local Demographers are using national high altitude black-and-white photography, combined with Space Shuttle Large Format Camera imagery, to develop techniques for estimating human population densities in urban areas (Lo 1989).

Regional Bird-watchers and waterfowl hunters are benefiting from the remote sensing efforts of the U.S. Fish and Wildlife Service and Ducks Unlimited. In central America and the prairie pothole region of North America, Landsat thematic mapper imagery is used to determine the number, size, and geographic distribution of wetlands important as breeding or overwintering areas for migratory waterfowl (Hill 1985). Mapping the location of waterfowl habitat is the first step toward effective management and protection.

National Use of the NOAA AVHRR sensor has generated continental-scale maps that monitor vegetation greenness (vigor) over time (Eidenshink 1992; Tucker et al. 1985, 1991; Prince 1991) and has facilitated development of both regional (Zhu and Evans 1992) and national forest cover maps.

Global Growing concern over the future of tropical forests has prompted an international effort involving CEC (Commission of the European Community), NASA/GSFC, UN/FAO, UNEP/GRID (Geneva). A CEC project called TREES (Tropical Ecosystem Environment observations by Satellites), operating out of the joint research center in Ispra, Italy, is coordinating a wall-to-wall AVHRR map (1 km resolution) of the entire tropical belt (TREES 1991). The map is designed as a baseline for long term monitoring.

during a hurricane would be difficult if not impossible to document by ground-based field surveys. But they are eminently practical using remote sensing imagery (plate 4). Timely use of remote sensing information has increased over the past twenty years, providing information essential for monitoring and mitigating environmental problems that occur suddenly and on a big scale.

Change is also of interest to environmental managers over much longer time frames. We may want to know, for example:

- How succession is proceeding within clearcut areas of the northwestern United States five, ten, and fifteen years after logging.
- What the average annual growing period of savanna grasses is in the Masai Mara region of southwestern Kenya.

Figure 3.1 Single frame of aerial videography. This single frame of aerial videography shows tree crown architecture well enough to identify not only the plant species but also the health of the individual tree.

- Whether near real-time vegetation monitoring can help predict the extent and severity of drought in eastern Mongolia.
- How far the southern edge of the Saharan Desert has advanced since the 1970s.
- How much the Aral Sea shrank in the past ten years.
- How dynamic the area of sea ice is in the northern Atlantic.

The ability to obtain data rapidly and inexpensively over large geographic regions means that remote sensing can help us document the local, regional, and global consequences of acute and chronic changes in ecosystems. Remote sensing thus can provide the knowledge essential for developing strategies to mitigate adverse impacts.

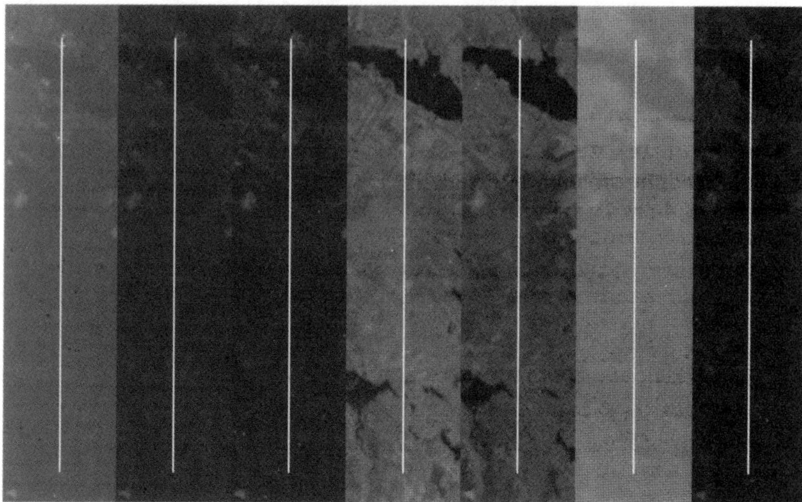

Figure 3.2 Spectral reflectance of the seven wavebands of Landsat TM imagery. The images are arranged sequentially with band 1 (blue) on the left. Notice how the detail (contrast) visible within each band varies. Infrared bands 4 and 5 have the most contrast (detail), partly because they are the least degraded by atmospheric effects such as haze. The white line within each image is one pixel wide and represents the transect of landscape brightness for each waveband shown on the line graphs. Notice how bands 4 and 5 show the greatest variability in brightness values (contrast), which correlates with the amount of detail evident in the images. Notice also that all reflected infrared wavebands (bands 4, 5, and 7) show a distinct drop in brightness around pixel 20, corresponding to an area of open water that strongly absorbs infrared radiation. The extremely bright spike in band 4 around pixel 151 is probably noise caused by a detector error.

Finally, in addition to its spatial and temporal advantages, remote sensing offers the advantage of a wide spectral coverage. For example, every autumn thousands of tourists visit New England to see the spectacular foliage. The brilliant red maples, sun-yellow aspens, and orange-tinged sumacs stand out in sharp contrast to the bluish-green backdrop of hemlocks and pines. Our eyes are capable of detecting thousands of shades of color (*hues*), and we have no difficulty distinguishing one tree species from another based solely on leaf color. Similarly, the red iron-oxide soils of Georgia look different from the sandy soils of Florida. And the same plowed field in Wisconsin will look very dark after a rain but much lighter after several days of hot sun.

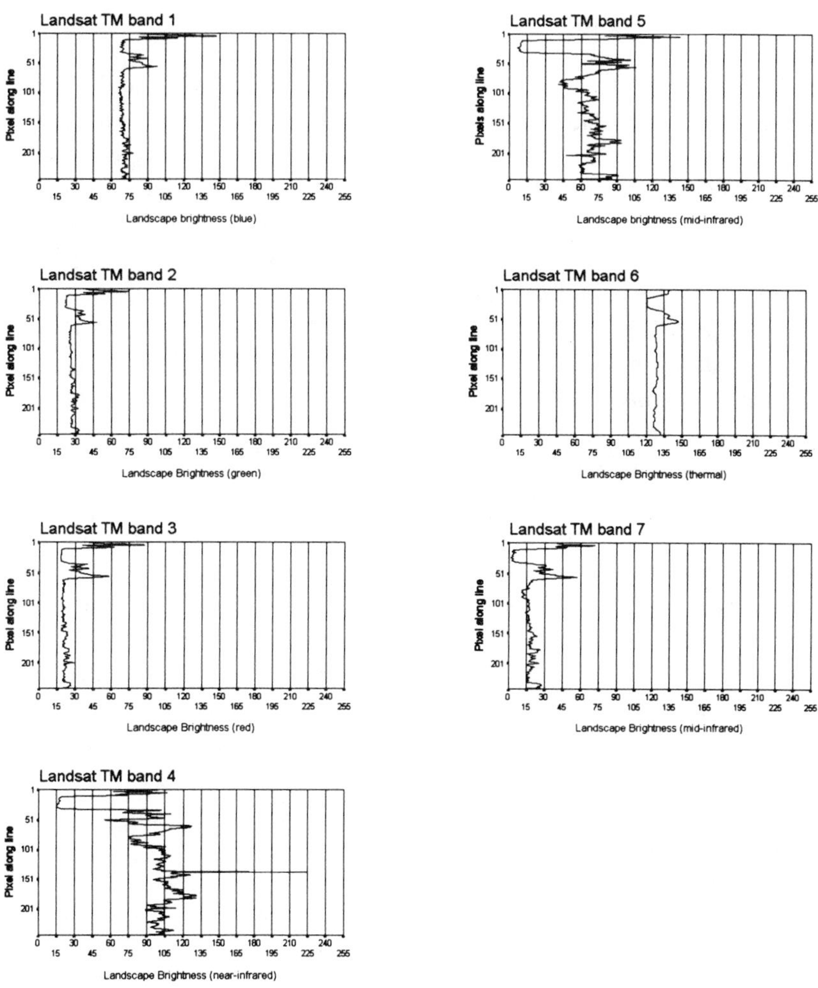

Because classes of landscape features, such as red maples, reflect radiation similarly, and because their color or *spectral signature* often contrasts with that of the background, we can use remote sensing to detect and identify objects within a landscape. Aerial photograph interpreters use infrared spectral characteristics to map the composition and distribution of conifer and hardwood woodlands within national forests (Heller et al. 1983) and to inventory and map wetlands throughout the United States (MacConnell et al. 1992). Figure 3.2, for example, shows the extent to which separate bands of blue, green, red, and infrared radiation are reflected and emitted from the same terrain features, such as water bodies, buildings, and woodlands.

Expert photointerpreters can distinguish 15–30 shades (tones) of gray on panchromatic (black-and-white) images, but can distinguish ten times as many hues on color infrared or true color imagery. Visual interpretation of imagery is limited by our ability to reliably discriminate the color or tone of a set of objects from that of all other objects in the background. Although our eyes are able to distinguish thousands of different hues when each appears on the same background color, accuracy declines dramatically when the background color varies—as it inevitably does in real life and on real terrain. The same is true when we attempt to distinguish different tones in a panchromatic image (figure 3.3). This reduces dramatically the effective number of colors and tones that a skilled interpreter is able to compare reliably within an image. The four bands of the Landsat MSS sensor, when combined, can represent more than 134 million ($2^{7+7+7+6}$; three spectral bands with 7 bit and one with 6 bit radiometric resolution) unique spectral reflectance values for any single *pixel* (picture element). A pixel is an element, or cell, within a *raster image,* which is a "picture" of the landscape made up of a rectangular mosaic of cells (pixels). The six reflectance bands of Landsat TM (thematic mapper) can generate 7×10^{16} reflectance values for each pixel. Visual interpreters can thus never even come close to making full use of the detailed spectral information available within multispectral remote sensing imagery.

Digital (statistical) interpretation of remote sensing imagery can, however, make full use of the range of colors recorded by multispectral sensors. One of the most powerful attributes of digital image processing is the capacity to statistically cluster pixels with similar spectral responses, thus isolating important features and effectively reducing the number of unique colors needed to create a visually informative map. The huge quantity of spectral information contained in multispectral remote sensing imagery allows researchers to differentiate features with very subtly different spectral responses (signatures).

Digital Landsat TM image analyses of the spectral response of Adelie penguin rookeries (composed of birds, guano, soil, and debris) relative to the background landscape in Antarctica allowed researchers to locate and map the extent of all rookeries on Ross and Beaufort Islands. By comparing the results to published estimates of penguin distribution, the researchers went on to identify previously undiscovered nesting sites

Figure 3.3 Classic illusion. Though the interior gray is the same tone in both triangles, it appears darker in the light bordered triangle. The same illusion applies to color combinations such as yellow against light and dark blue. This illusion can make it difficult to identify the same color feature within an image when the contrast of the background varies.

(Schwaller et al. 1989). The same principle was used to map the size and location of near-surface schooling Pacific herring with experimental airborne spectrographic imagery (Borstad et al. 1992).

How Much Detail Is Achievable and Necessary?

Landscapes are composed of unique features, which should be thought of as useful information rather than as physical objects on the earth's surface. These features are arranged either as a *mosaic* that completely covers the area (e.g., an urban landscape composed of buildings, parking lots, parks, etc.) or as discrete elements distributed on a continuous background (e.g., patches of vegetation against a desert soil). Landscapes can be simple, composed of one element and the background, or complex, composed of many elements (Woodcock and Strahler 1987).

Selecting an appropriate source of remote sensing data requires that we understand how the resolution of the system will affect our ability to measure the biophysical landscape features that are of interest. The resolution of a remote sensing device, from a hand-held camera to a satellite-based sensor, is a measure of the smallest feature (a physical object or event) that can be distinguished spatially, spectrally, temporally, or radiometrically from other features taken as background. Whether a feature can be detected and identified depends on several factors. Two of the most important are (1) the field-of-view of the remote sensor and (2) the contrast ratio between the terrain feature of interest and the background.

Spatial Resolution

The minimum size of terrain features that can be distinguished from the background or whose dimensions can be measured determines the spatial resolution. With aerial photography, spatial resolution is a function of the grain of the film (measured in line pairs per mm that can be resolved) and the scale of the photography. The latter is a function of the focal length of the camera and the altitude of the aircraft (figure 3.4). A very fine grain film will resolve more detail than will a coarse grain film at the same scale. Regardless of the grain of the film, the larger the scale, the higher will be the resolution of the image.

Not all aerial photography is undertaken from aircraft. The Large Format Camera (LFC) and Metric camera (a standard aerial photographic camera—Zeiss RMK A30/23 mounted in a pressurized container) were both operated on an experimental basis during space shuttle flights (Slater et al. 1983). At an altitude of 250km the LFC was able, using black and white film, to image 190×190km areas at a scale of 1:820,000 and a spatial resolution of approximately 20m.

For nonphotographic sensors spatial resolution is determined in several ways. The most common is the dimension of the ground projected instantaneous field-of-view (IFOV) of the system, which is a function of the size of the detector and the sensor altitude and optics. For example, the SPOT multispectral satellite sensor has a nominal IFOV of approximately 20×20m (figure 3.5), such that as the sensor passes over a section of the earth it records the amount, or *brightness,* of radiation reflected and emitted from a continuous series of $400m^2$ areas or picture elements (pixels).

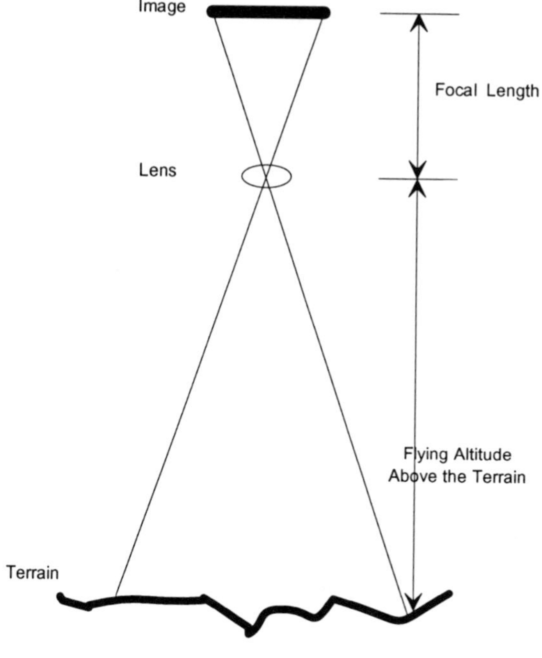

Figure 3.4 Scale as a function of focal length and altitude. In this example, an image taken with a 35mm camera flying at an altitude of 3,500m above the surface would have a scale of 1:100,000.

The actual IFOV varies because no satellite has a perfectly stable orbit. Landsats 1 to 3 had an average altitude of 913km with a range from 880–940km. As a satellite drops in altitude, its IFOV will become smaller. The IFOV of the Landsat MSS system is 79×79m (6,241 m^2), but actual resolution varies from 76×76m to 81×81m.

It is important to understand that digital sensors such as those on the Landsat and SPOT satellites record the brightness of all features that lie within their IFOV. *Brightness* is the amount of radiation reflected or emitted from landscape features. Objects that reflect most of the incident sunlight will appear bright in the remote sensing images. The digital value

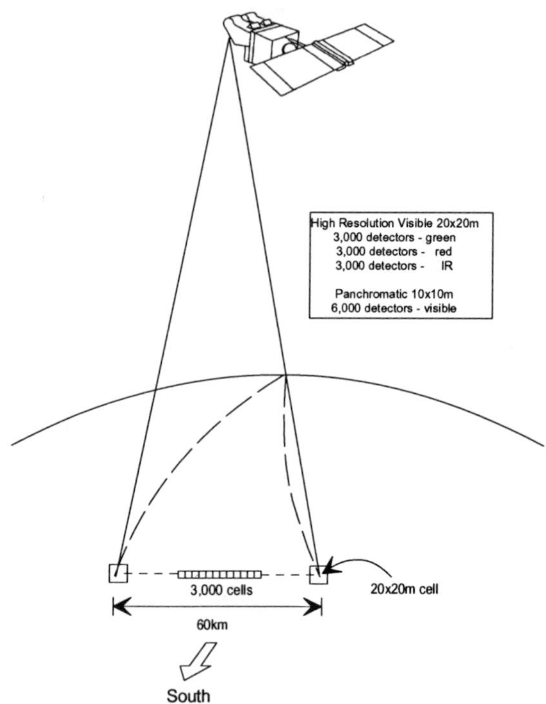

Figure 3.5 Resolution and width of SPOT imagery. The SPOT near-polar orbiting satellite carries two charges coupled device array sensor systems. The High Resolution Visible (HRV) multispectral sensor records terrain brightness within green, red, and IR wavebands at a spatial resolution of 20m; the panchromatic sensor records terrain brightness in one broad waveband covering all visible light at a spatial resolution of 10m. The SPOT detector arrays view a 60km wide swath of terrain at any one time, and store the radiation reflected from the landscape as a line of 3,000 pixels for HRV and 6,000 pixels for panchromatic.

recorded for a given pixel is the sum of the brightness values of all features within the IFOV.

The IFOV (instantaneous field-of-view) is the area of terrain within which the composite brightness of features is measured. The *brightness value* of a pixel is derived from raw IFOV brightness values. However, the size of a pixel can be larger or smaller than the IFOV of the sensor, depending on how the IFOV brightness values are sampled (recorded).

As IFOV brightness is a total feature brightness, if all features have similar reflectance, then the feature that covers the largest area of the IFOV will contribute the most to the recorded IFOV brightness. If, however, a small but relatively bright or dark (contrasty) feature exists within the IFOV, then its high contrast will skew the total brightness value. That feature will thus contribute proportionally more to the composite spectral response of features within the IFOV (figure 3.6). The fact that contrasty features contribute more to the composite brightness of an IFOV than does the background explains why some features (e.g., roads within agricultural fields) that are smaller than the IFOV of a sensor can still be detected within the imagery (figure 3.7).

It is very important to emphasize that for the spatial resolution of a remote sensing system to be appropriate to the question, the feature of

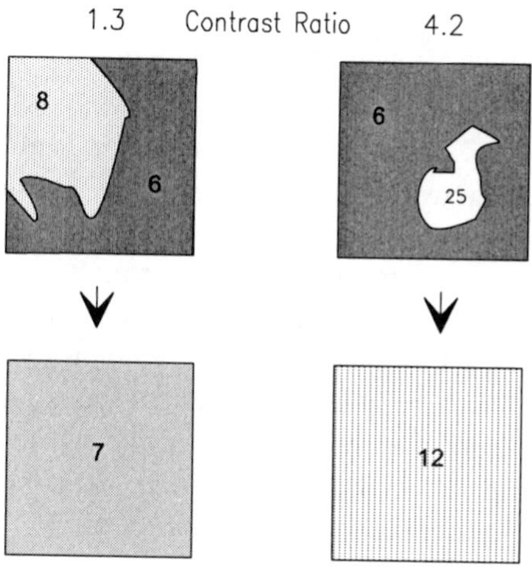

Figure 3.6 Contrast effects on composite radiance. The brightness value of a pixel is based on the composite brightness of all landscape features contained within the pixel. Small bright or dark objects can contribute disproportionately to the overall brightness of the pixel, as is shown in the example on the right. High contrast ratio of a subpixel-size feature (e.g., narrow roads through a wheat field) may bias the brightness value of pixels, allowing the feature to be detected within an image.

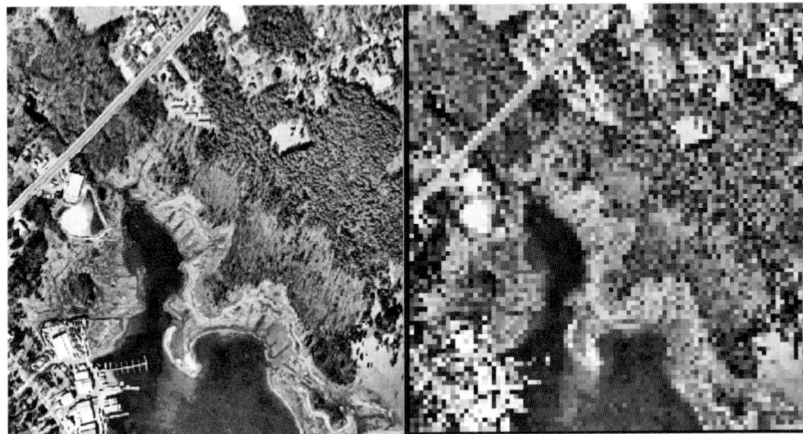

Figure 3.7 Resolving subpixel-size features. *Left:* high resolution image with contrasty target. *Right:* degraded resolution image, which still shows the subpixel-size features. Note how the pier projecting into the water in the high resolution image (lower left corner) is still detectable in the low resolution image even though it has become a subpixel-size feature.

interest must not only be detectable by the sensor. It must be identifiable and analyzable by the researcher. *Detectability* is the ability of a remote sensing system to record the presence of absence or a feature on the landscape. A feature can be detected even if it is smaller than the theoretical spatial resolution of the sensor (e.g., a high-contrast object such as a concrete road through a wheat field). *Recognizability* is the ability of the human interpreter to identify (put a name to) a feature detected by the sensor systems. A feature may be detected by the sensor but may not be recognizable (e.g., narrow straight lines in an image may be roads, railways, or canals). Recognizability is a function of interpreter experience and image scale.

For example, while aerial photographic images obtained over the plains of southern Sudan may show small circular features that are clearly different from the background savanna, there is insufficient detail for the researcher to identify the features as the camps of cattle herders or as patches of grewia shrubs. And only if detail (spatial resolution) were increased considerably could the researcher analyze the camps to determine if they were occupied or not.

A general rule of thumb regarding spatial resolution is as follows: To move from detection to identification requires a threefold increase in resolution; to pass from identification to analysis may require a tenfold or greater improvement in resolution. For example, if an image from Landsat TM (30m resolution) was being used for a study of waterfowl breeding areas in the prairie pothole region of South Dakota, then lakes and ponds would have to exceed 0.5ha (5 pixels) for detection, 1.5ha (15 pixels) for identification (e.g., vernal or permanent pond), and 10ha (100 pixels) for analysis (e.g., eutrophic status).

The spatial resolution of a remote sensing system should be less than half the smallest dimension of the feature to be detected. When the spatial resolution is coarser than this, individual pixels are composed not of spectral reflectance from a unique feature of the landscape but from a melange of spectral reflectances from a mixed set of features (figure 3.8). Mixed pixels will of course exist along the edges of objects regardless of the spatial resolution (figure 3.9). Thus, when determining the spatial resolution required to observe a specific phenomenon, a researcher must be aware that some portion of the feature will be composed of mixed pixels, and therefore will be ambiguous.

For a unique feature to be both detected and identified successfully it should be made up of no fewer than 3–5 *pure pixels*—that is, pixels composed solely of the unique reflectance from the feature of interest (with no edge effects). For many agriculture and forestry applications this translates to feature sizes being greater than 20–50 IFOVs or pixels (Jensen 1986). It is important to remember that a pure pixel is user defined and may be composed of a range of terrain features. For example, a pure pixel for a forest is actually a composite of the reflectance of all tree species' canopies and the visible midstory and understory. Similarly, though a suburban pixel may be considered pure if it is composed of houses, streets, and gardens, it may be considered *mixed* if it contained agricultural land or a rail yard. Table 3.2 provides a very general guide to match categories of applications with the appropriate spatial resolutions of sensor systems.

Higher spatial resolution, perhaps surprisingly, does not always produce more useful information. The reason is related to the maxim "not seeing the forest for the trees." For example, we may be able to detect a landscape of mixed wooded savanna using 80m resolution Landsat MSS imagery and thus contrast it to pure forest and pure grassland. Yet, if

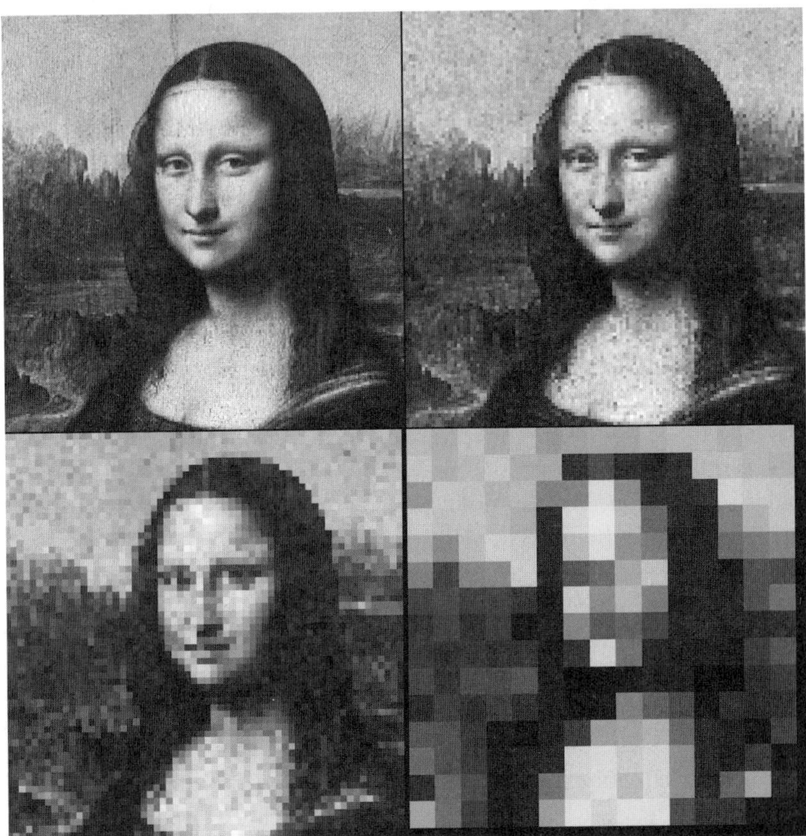

Figure 3.8 Identification versus detection. This series of images of *The Mona Lisa* shows how, as the spatial resolution of the image is gradually degraded, features become unidentifiable well before they become undetectable. Note how the smile is still detected in the lower left image but is not identifiably enigmatic. In the very coarse resolution image at the lower right, the hair and face are detected (i.e., they have different brightness values compared to the background), but they are not identifiable as such.

we increased the resolution by using 10m SPOT panchromatic imagery, individual pixels within wooded savanna areas might now be classified as either trees or grasses. Although objectively accurate, the higher resolution of the data actually reduces the useful information content of the image by making it harder to distinguish mixed wooded savanna from

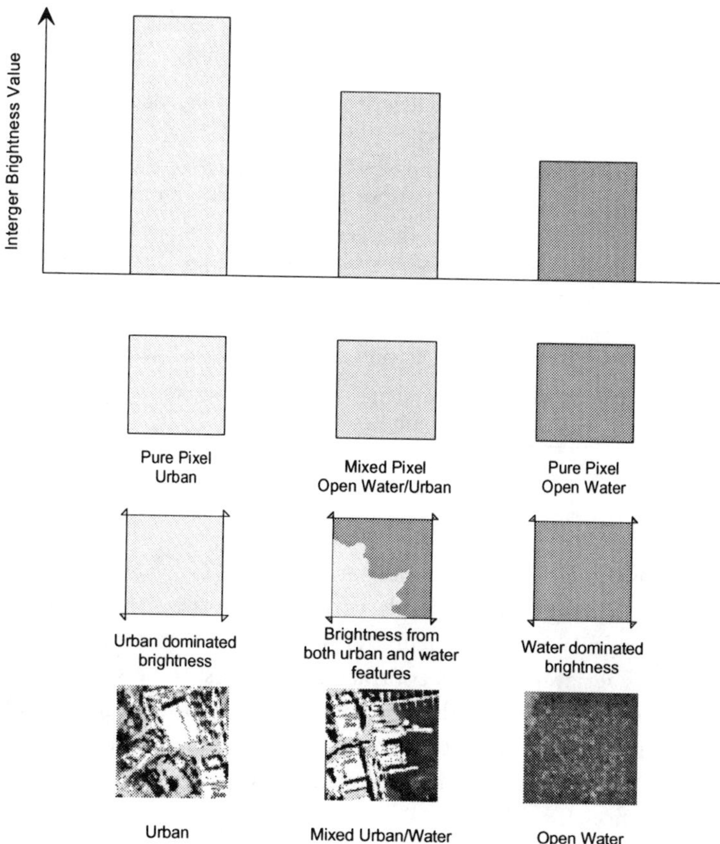

Figure 3.9 Brightness values of pure and mixed pixels. The integer value stored within a pixel is an aggregate of the brightness of all features within the pixel. The middle pixel is approximately a 50/50 mixture of urban landscape and open water. As a result its brightness value falls between that of the pure pixels, and thus does not belong to the set of pixels classed as either urban or open water. In very heterogeneous landscapes the majority of pixels in an image may be mixed.

Table 3.2
Features Identifiable at Different Scales of Imagery

Scale	Identifiable Features
1:500	Plant species identification, size of individual trees, uses of buildings, functions of industries
1:5,000	Volume of timber, wetland boundaries, outline minor tributaries, transportation networks, property boundaries
1:50,000	Outline areas of evergreen and deciduous trees, outline areas of forest associations, determine direction of flow of water, outline shorelines, locate major transportation routes, measure areas of agricultural land
1:500,000	Regional vegetation and land use classification
1:5,000,000	Major river systems, continental vegetation zones, continental cloud cover

forest or grasslands. The optimum spatial resolution for any given study will be that which generates pixels with a composite spectral response that is meaningful for the particular classification scheme being used. In reality, unless aerial imaging is contracted to meet specific spatial resolution requirements, users usually have only a very narrow range of spatial resolution imagery available to select from (i.e., public domain or commercially obtained aerial photography and satellite imagery at predetermined resolutions).

Spectral Resolution

Spectral resolution refers to the dimension and number of wavelength regions in the electromagnetic spectrum to which the sensor or recorder is sensitive. Film for black-and-white aerial photography has a broad spectral resolution, in that the silver emulsion is sensitive to all reflected blue, green, and red light between 0.4 and 0.7μm. In contrast, band 3 of the Landsat TM sensor has a relatively fine spectral resolution, recording radiation only between 0.63 and 0.69μm. When sensors with a large number of narrow bands (>200 bands of 0.01μm width), such as the Airborne Visible and Infrared Imaging Spectrometer (AVIRIS), become widely available, a researcher will then be able to select those bands that maximize the spectral contrast between the feature of interest and the background. This will improve the likelihood that the feature not only will be detected but can be identified and analyzed.

For example, if you know the typical spectral response of pine and fir trees, then selection of spectral bands at 0.55μm, 0.7μm and 0.9μm would be most likely to allow differentiation of the two species (figure 3.10). If, of course, the leaves of two tree species show little difference in spectral signature, then you would only be able to differentiate between them if spatial resolution were sufficient to detect differences in architecture, or if temporal resolution allowed detection of phenological differences (i.e., flowering times).

Radiometric Resolution

Radiometric resolution is a measure of the sensitivity of a sensor to differences in the intensity of the flow of radiation (radiant flux) reflected or emitted from features on the ground. For example, as radiation between

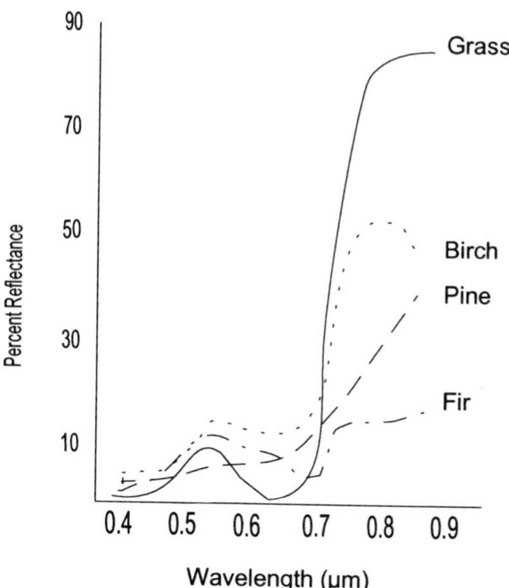

Figure 3.10 Typical spectral reflectance of grass, birch, pine, and fir trees.

wavelengths 0.6 and 0.7μm hits the band 5 Landsat MSS detector, a voltage is generated. This analog (real) voltage is sampled every 9.958×10^{-6} seconds, and rounded and scaled to a single (integer) value ranging from 0 to 127, giving 128 discrete levels (7 bits) of radiation intensity (figure 3.11). When this happens, subtle differences in radiance values can be lost, causing unique features on the ground to be recorded in the imagery with the same digital reflectance value (brightness).

At a given spatial and spectral resolution, increasing the radiometric resolution is likely to increase the likelihood of detecting and identifying features of interest. For example, even if the spatial and spectral resolution of an image provides sufficient information to detect forest patches along a stream channel, the radiometric resolution of the MSS sensor (64 brightness values; rescaled to 128 levels for bands 4, 5, and 6) may be too coarse to identify the differences between the spectral signatures of willows and swamp oak. The Landsat TM sensor records radiation brightness with 8 bit resolution (0 to 255 levels), providing much greater feature discriminating power than the older MSS sensor system, and thus may have a greater likelihood of differentiating these two bottomland hardwoods.

Increasing the radiometric resolution of a sensor system requires that the difference between one brightness level and the next is greater than the noise level of the system. Otherwise it would be unclear whether differences in brightness levels for adjacent pixels were a result of true differences in feature reflectance or merely errors associated with the sensing system. The new pushbroom type of sensor that uses a linear array of individual detectors (rather than a single detector and a scanning mirror) spends longer looking at each point (dwell time) and thus has a higher signal-to-noise ratio. Pushbroom systems, such as that used in the SPOT satellites, are capable, therefore, of higher radiometric and spectral resolutions than are the older scanning mirror systems (Norwood and Lansing 1983). At present, no pushbroom system in orbit exhibits a higher radiometric resolution than do the scanning radiometer systems. In the case of the SPOT system, because of a data transmission constraint, designers traded off better radiometric resolution for better spatial resolution.

Temporal Resolution

Temporal resolution is a measure of how often a given sensor system obtains imagery of a particular area (i.e., how often an area can be revisited).

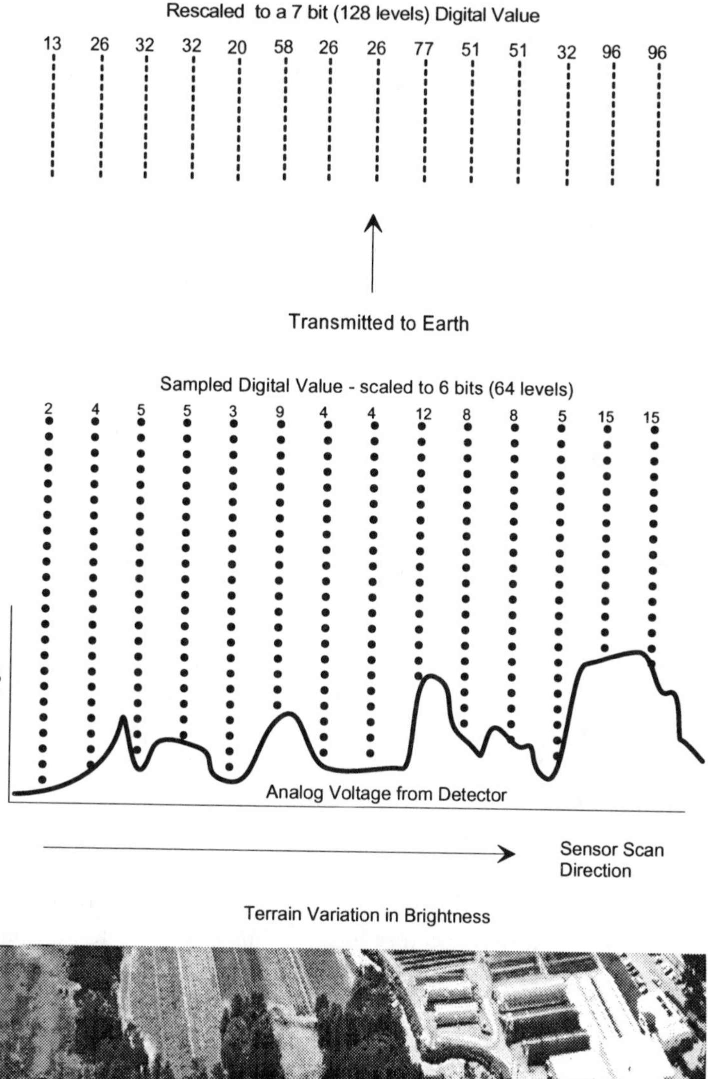

Figure 3.11 How a sensor generates a digital representation of terrain brightness. As the sensor scans across the terrain from left to right, radiation is focused on the detector, generating a voltage proportional to the terrain brightness. During the sensor scan the continuous (analog) detector voltage is measured (sampled) at regular time intervals. The sampled voltage is converted to a 6 bit integer value (i.e., has a range from 0–63). This brightness integer is transmitted to earth and, in this example, is rescaled to 128 brightness levels (7 bits) to generate greater contrast in the final image.

This simple definition obscures how important temporal resolution can be as an aid to detection, identification, and analysis of features of interest.

There are few biological or physical features or phenomena that do not change over time: deciduous trees gain and lose their leaves; annual crops flower, seed, and senesce; urban landscapes expand or change in land use; riverbanks flood and dry out; abandoned agricultural areas become fallow fields, shrublands, and perhaps forests. For many of these physical or cultural features there are optimal times during which each is most evident. A time-varying characteristic thus can often be used when spatial, spectral, or radiometric resolution is insufficient for detection. For example, wooded wetlands and hardwood forests are often difficult to distinguish in aerial photographs if the imagery is obtained when trees are in leaf. If, however, imagery is obtained before leaf-flush in the spring when wetlands are still waterlogged, the observer not only can see the hardwood trees but can identify which are growing in flooded or in upland areas.

For many applications, the timing of image acquisition plays an important part in detection, identification, and analyses of relevant features. For example, two grain crops may appear very similar spectrally, and are thus indistinguishable for much of their vegetative growing period. If, however, they flower at different times, timely acquisition of remote sensing information may allow a knowledgeable observer to distinguish the flowering from the nonflowering crop, without the need for higher spatial or spectral resolution imagery. In many cases like this, effective use of the temporal nature of a sensor can reduce spatial, spectral, and radiometric resolution needs, thus lowering the cost of data acquisition and processing.

For short duration or catastrophic phenomena, remote sensing imagery can provide valuable information over large areas, which would be almost impossible to obtain by on-the-ground means. Remote sensing imagery has assisted greatly in the assessment of hurricane and flood damage (Barton and Bathols 1989; Blasco et al. 1992; Brown et al. 1987), as well as pollution dispersion (Cross 1992; Spitzer et al. 1990) and effluent discharge (Delregno and Atkinson 1988). Some satellite-based remote sensing systems, such as the NOAA AVHRR, allow for daily or twice daily images to be obtained for any area of the globe. Such high temporal resolution allows researchers to track rapidly changing phenomena, such as floods (Barton and Bathols 1989) or the greening and senescence of seasonally arid savannas (Achard and Blasco 1990). Daily monitoring of vegetation greenness within a region of the Sahel would reveal the onset and progress of the growing season and thus help guide the selection of appropriate plants for soil stabilization and afforestation (Prince 1991).

Resolution Trade-Offs

Most first-time users of remote sensing imagery assume that higher resolution provides more detail which, in turn, must yield more information for feature identification and analyses. Although it is true that higher resolution generates more data, more data is not always synonymous with more information. Figure 3.12 shows 512×512 pixel black-and-white images of *The Mona Lisa.* The top left rendition uses 256 shades of gray (brightness levels), whereas the middle right uses only 16 gray tones. Surprisingly, the human eye perceives both images as having comparable levels of detail. The additional data in the 256 gray-scale image provide us with little extra visual information.

In some instances, more detail merely introduces noise that obscures the features of most interest to the observer. For example, if we are interested in distinguishing farmland from urban areas, we do not need to be able to discern every grassy patch in order to determine whether the area is urban or farmland. Very high spatial detail may actually increase confusion by detecting urban lawns, gardens, and parks that may be identified erroneously as farmland. Thus, in this case, lower resolution imagery would be more effective in discriminating between urban and farmland areas.

From a technical point of view, there is always a trade-off between resolution (spatial, spectral, radiometric, and temporal) and the cost of image acquisition and processing. Increasing resolution causes a parallel and multiplicative increase in the quantity of data that has to be obtained, stored, and analyzed. Complete coverage of Massachusetts at a scale of 1:12,000 requires 11,556 aerial photographs. At 1:25,000 scale the requisite photographs drop to 2,889, while at 1:60,000 scale only 300 photographs are needed. The costs of purchasing and developing the film, let alone the cost of analyzing the imagery, rise dramatically as resolution increases. For example, a Massachusetts statewide wetland inventory, using aerial photography, would cost over $8 million and take 200,000

Figure 3.12 Diminishing returns for higher resolution imagery. Our eyes, sensitive to light in the blue-green-red range 0.4–0.7μm, are able to distinguish hundreds of thousands of colors (hues), but we are capable of detecting only about 20–30 levels of brightness. This black-and-white series of images of the *Mona Lisa* is displayed with 2 (*lower right*), 4, 16, 32, 64, and 256 (*upper left*) gray tones. Note that, to the human eye, there is little to gain (informationally) from increasing the resolution beyond 16 shades of gray.

man-hours to complete at a scale of 1:5,000. The same inventory would cost only $1 million and require only 30,000 man-hours at a scale of 1:12,000—with no loss of accuracy (MacConnell et al. 1992).

A Landsat MSS image covering 185×170km, with a spatial resolution of 56×79m, 4 spectral channels, and a radiometric resolution of 6 bits (64 levels) for band 7 and of 7 bits (128 levels) for bands 4, 5, and 6, requires 24Mb (megabytes = million bytes) of disk or tape storage. In contrast, a Landsat TM image covering the same area, with a spatial resolution of 30×30m, 7 spectral channels, and a radiometric resolution of 8 bits (256 levels) requires 227Mb of storage. As resolution increases, data storage becomes a major constraint. A trade-off must be made between detail and area coverage. If a personal computer monitor is capable of displaying 1,024×768 pixels, a single screen of Landsat MSS imagery resampled to 79×79m resolution covers some 81×61km (4,941km^2), whereas a single screen of SPOT panchromatic 10m resolution data covers only 10×7km. The Landsat MSS data provide a single synoptic view of an area, whereas the SPOT data of the same area must be viewed in sections—each of which provides greater spatial detail (figure 3.13). Users of remote sensing information must resolve the competing goals of viewing large areas and detecting small features. To map the area coverage of rain forest in the Central African Basin would require 15,000 aerial photographs at a scale of 1:90,000, a hundred Landsat TM scenes, but only two AVHRR scenes.

In light of the trade-offs in data acquisition, storage, and processing associated with increased resolution, it is exceedingly important that the choice of image resolution be problem driven. Researchers need to ask themselves: What features do I need to detect, identify, and analyze? What is the optimum mix of spatial, spectral, radiometric, and temporal resolutions needed to satisfy my objectives within the constraints of time, money, and manpower?

Choosing the Appropriate Imagery

Remotely sensed information can be drawn from any of these sources: multispectral and radar imagery, aerial photography and videography. How do you choose which will best meet your needs? What are the advantages and disadvantages of each sensor system?

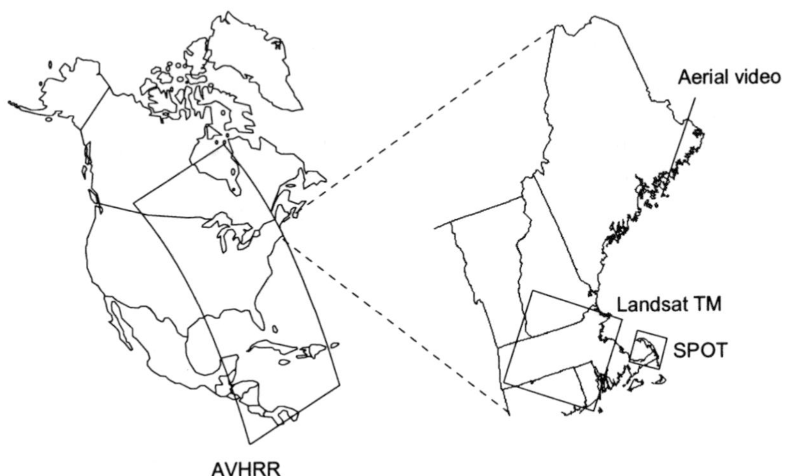

Figure 3.13 Spatial coverage of single images obtained by four different remote sensing systems.

Satellite imagery Satellite imagery can be purchased from commercial companies (e.g., EOSAT, SPOT Image) or from government agencies (e.g., United States Geological Survey, Eurimage). Images can be obtained in digital form (i.e., images on tape or disk must be transferred to, and viewed on, a computer) or as black-and-white or color prints and transparencies. Although imagery from reconnaisance satellites is available from 1960 (McDonald 1995), global coverage is not available prior to the 1972 launch of Landsat. Satellite imagery is most appropriate for applications that

- demand a regional or global perspective (at least 10,000km^2);
- require several spatially separate areas to be surveyed, monitored, or compared;
- require frequent, repetitive coverage;
- require or can take advantage of multispectral data; or
- target features larger than the spatial resolution of the imagery.

Advantages of satellite imagery are

☺ a wide spectral range (UV–IR);
☺ amenable to both digital and visual analysis;
☺ availability of both digital and photographic (analog) output;

- ☺ a wide dynamic range (brightness values) of detectors;
- ☺ quantitative biophysical measurements from radiometric information obtained from calibrated sensors;
- ☺ a consistent sun angle, perspective, and sensor response;
- ☺ suitable for comparing different scales and wavelengths; and
- ☺ semiautomated processing that makes use of full dynamic range of the data.

Disadvantages of satellite imagery are

- ☹ low spatial resolution relative to airborne and ground-based sensors; and
- ☹ higher start-up equipment and training costs.

Large format aerial photography Large format aerial photography (i.e., at least 70mm filmstock) is used in most countries throughout the world to generate topographic and land use maps. Thus archival panchromatic, color, and IR aerial photography can often be purchased from national or state cartographic agencies. If photographs are not available for your area of concern, commercial companies can be contracted to obtain photographs at the most appropriate time and scale. Large format aerial photography is most appropriate for applications that

- demand a small area or local perspective (less than 1,000km^2);
- constitute one-time or baseline mapping; and
- target small features (less than 10m in diameter).

Advantages of aerial photography are

- ☺ superior spatial resolution;
- ☺ little geographic rectification required;
- ☺ simple operation; and
- ☺ low cost of analysis equipment.

Disadvantages of aerial photography are

- ☹ limited spectral range photographic film;
- ☹ digital analysis obtained only by scanning photographic prints or transparencies;
- ☹ difficulty in interpreting large volumes of data (large areas); and
- ☹ data acquisition and analysis costs high per km^2 for areas greater than 1,000km^2.

Aerial 35mm photography Aerial 35mm photography is most often obtained on an ad hoc basis by the researcher. Using your own 35mm camera, oblique photographs are shot through the open windows of a rented small plane flying over the target area. Aerial photography is most appropriate for applications that

- involve one or a few small areas (less than 5km^2);
- require frequent, repeated coverage; and
- target very small features (less than 5m in diameter).

Advantages of 35mm aerial photography are

☺ very inexpensive for micro-scale surveys (sampling) and monitoring (aircraft rental $100–$200 per hour);
☺ superior spatial resolution (a scale of 1:500 or better is possible);
☺ simple operation; and
☺ low cost of analysis equipment.

Disadvantages of 35mm aerial photography are

☹ oblique photographs make it difficult to estimate the area of features and to mosaic photographs together;
☹ digital analysis is obtained only by scanning photographic prints or transparencies;
☹ few photoshops develop color infrared film;
☹ photographic film has limited spectral range; and
☹ large volumes of data (large areas) are difficult to interpret.

Aerial videography Aerial videography can be obtained much like 35mm photography by pointing a videocamera or camcorder through the window of rented airplane. More often, however, the videocamera is mounted outside the window pointing directly down, and it is controlled remotely by the researcher within the plane. Aerial videography is most appropriate for applications that

- involve a few relatively small areas (less than 50km^2) or samples along a transect or linear feature;
- require frequent, repeated coverage; and
- target relatively large features (at least 25m in diameter).

Advantages of aerial videography are

- ☺ least expensive system for small- to mid-scale surveys (sampling) and monitoring;
- ☺ ability to view imagery during image acquisition;
- ☺ visual and digital analysis possible;
- ☺ simple operation;
- ☺ visible to near-IR spectral range of video cameras;
- ☺ low cost of acquisition and analysis equipment; and
- ☺ able to annotate key landscape features using the audio channel of the videotape.

Disadvantages of aerial videography are

- ☹ low spatial resolution;
- ☹ requires relatively high light levels for image acquisition; and
- ☹ removal of spatial distortion required for area estimation of features.

Radar Radar imagery can be obtained either by plane or satellite. The preceding remote sensing systems all depend on the sun to illuminate landscape features. In contrast, radar systems provide their own source of illumination by transmitting an EMR signal in the microwave region. These microwaves are reflected back from landscape features and are detected by the sensor. The timing and intensity of the return signal is used to generate the final radar image. Radar imagery is most appropriate for applications that

- involve areas of perennial cloud cover; and
- target relatively large features, (at least 50m in diameter).

Radar has one great advantage:

- ☺ ability to obtain data regardless of cloud cover or weather conditions.

Disadvantages of radar are:

- ☹ the high cost of airborne equipment;
- ☹ satellite systems primarily experimental;
- ☹ low spatial resolution;
- ☹ image analysis and interpretation methods for natural resource management not well developed; and
- ☹ provides information on terrain texture and water content only.

Table 3.3
Image Processing Hardware and Software Configurations

A basic IBM PC—compatible image processing system		
Pentium 75Mhz CPU system; 8Mb RAM, 1Gb hard-disk, 1.44Mb diskette drive, and 1024 resolution 14-inch 16 bit color display.		$2,000
Color inkjet printer 9×11".		$800
Tape backup 350Mb.		$260
An advanced IBM PC–compatible image processing system		
Pentium 120Mhz CPU system; 16Mb RAM, 1Gb hard-disk, 1.44Mb diskette drive, and 1024 resolution 20-inch 24 bit color display.		$5,000
A Macintosh-compatible image processing system		
PowerMac 8100AV 10MHz PowerPC CPU, 16Mb RAM, 1.0Gb hard-disk, 1.44Mb diskette drive, and 1024 resolution 21-inch 24 bit color display, CD-ROM..		$5,500
A UNIX workstation–based image processing system		
SPARC station 20, 32Mb RAM, 1.0Gb hard-disk, and 1024 × 768, 21-inch 24 bit color display, CD-ROM		$13,000
Peripherals for advanced PC, Macintosh, and UNIX systems		
Digitizing tablet, 3'×4'.		$1,500
Color postscript printer (inkjet).		$1,500
Dye sublimation printer (600dpi).		$2,500
Color laser printer.		$5,000
Color (24 bit) 800dpi scanner.		$700
8 mm backup Tape Drive 5-14Gb.		$2,500
IBM DOS Image processing and GIS software		
ArcView 1.0	http://www.esi.com.	free
ArcView 2.1	Fax: (909) 793-5953.	$500
IDRISI	Fax: (508)793-8842..	$500
Dragon	Fax: (413)549-6401..	$1,000
TNTmips	Fax: (402)477-9559..	$4,000
PCI EASI/PACE	Fax: (703)243-3700..	$5,500
Macintosh Image processing and GIS software		
ArcView 21	Fax: (909) 793-5953.	$500
DIMPLE	Fax: (617)739-4836..	$2,000
TNTmips	Fax: (402)477-9559..	$4,500
MultiSpec	Landgreb@ecn.purdue.edu	Shareware
DRIGO	Fax: (805) 893-8617.	$50
UNIX Image processing and GIS software		
ArcView 2.1	Fax: (909) 793-5953.	$500
ERDAS	Fax: (404)248-9400..	$15,000
ARC/INFO	Fax: (909)793-5953..	$15,000
TNTmips	Fax: (402)477-9559..	$10,000
PCI EASI/PACE	Fax: (703)243-3700..	$10,000

NOTE: Prices as of August, 1995.

Table 3.4

Imagery Acquisition and Analysis Equipment Costs

IMAGERY	ACQUISITION		MEDIA		ANALYSIS EQUIPMENT	
Aerial Photography					see table 3.3	
35 mm	Aircraft rental	$150/hr	35 mm print or slide film	$20/36exp	Basic digital analysis system	
	35 mm camera	$600-$2500	35 mm color IR	$40/36exp	color scanner	$800-$1500
Digital	Aircraft rental	$150/hr			Advanced digital system	
	Digital camera	$600-$10,000				
Videography	Aircraft rental	$150/hr	Video-8 tapes	$5-$10	Basic digital analysis system	
	Hi8 color camera	$900-$2000			video-frame grabber	$400-$700
	GPS with data port	$800-$3000				
	Character generator	$3500-$4500				
Commercial 70 mm			1:5,000		**Visual**	
			Panchromatic	$15-$20/print	Stereoscope	$30-$500
			Color IR	$15-$20/print	Zoom transfer scope	$15,000
			Analog orthophotos	$50-100/print	**Digital**	
			1:60,000		Basic digital analysis system	
			Panchromatic	$200-$300/print	color scanner	$800-$1500
Satellite Imagery						
SPOT			1:200,000 Panchromatic	$1800/print		
			1:100,000 Panchromatic	$1800/print		
			Digital tapes 60 x 60 km scene	$2,500	Advanced digital system	$5,000
			Geocorrected digital tapes	$3,000	Advanced digital system	$5,000
Landsat TM			1:1,000,000	$2700/print		
			1:100,000 100 x 100km area	$2900/print		
			Digital tapes 180 x 170 km	$4,400	Advanced digital system	$5,000
Landsat MSS			Digital tapes 180 x 170 km	$1,000	Advanced digital system	$5,000
			2+ years old digital tapes	$200		
NOAA AVHRR			Digital tapes 2700km swath	$90	Advanced digital system	$5,000

Comparative Costs of Using Remote Sensing

As could be expected, the costs of remote sensing mapping, surveying, and monitoring increase with increasing spatial resolution, area coverage, and frequency of coverage. Choosing a cost-effective method of obtaining and analyzing remote sensing information cannot simply be based, however, on finding the least-cost method per unit area. Start-up costs in terms of equipment and personnel training may make, for example, a one-time digital image survey prohibitively expensive, while making multiple surveys exceedingly cost effective.

When selecting a remote sensing analysis system for the first time, it is important to consider the likelihood of undertaking a similar project or problem in the future. If repeated need is forecast, one must try to choose a system that can meet the present need and yet be adapted or expanded easily to meet future needs. Because of rapid advancement in computer hardware and software technology, first time users of digital remote sensing imagery should not necessarily opt for the most complex, full-featured, and thus expensive system. Rather, we would suggest selecting an entry-level system that can address the specific problems that motivated the use of remote sensing analysis, and that can be expanded if the user finds remote sensing analysis a useful tool. This avoids the all too common occurrence of purchasing a complex and expensive tool in the rarely met expectation that projects will materialize to keep it in use.

4

Commonly Used Types of Remote Sensing Data

The information contained within remote sensing data differs primarily according to the sensor used to record the electromagnetic radiation reflected or emitted from the landscape, and the platform used to position the sensor at some distance from the terrain.

Overview of Platforms and Imagery Sources

Platforms that carry the sensors that gather remote sensing information can be grouped according to whether they are terrestrial, airborne, low orbiting, or geosynchronous orbiting (figure 4.1). Terrestrial platforms range from simple tripods that raise the sensor 1–2 meters above the object to be sensed, to booms and cranes and towers that can reach tens of meters above the surface. Airborne platforms (including airplanes, helicopters, balloons, and even rockets) operate from sea level to well into the stratosphere at an altitude of about 500,000 meters. Low-orbiting (200–950km) platforms are either short duration, such as the space shuttle that remains aloft for 1–2 weeks, or long duration, such as the

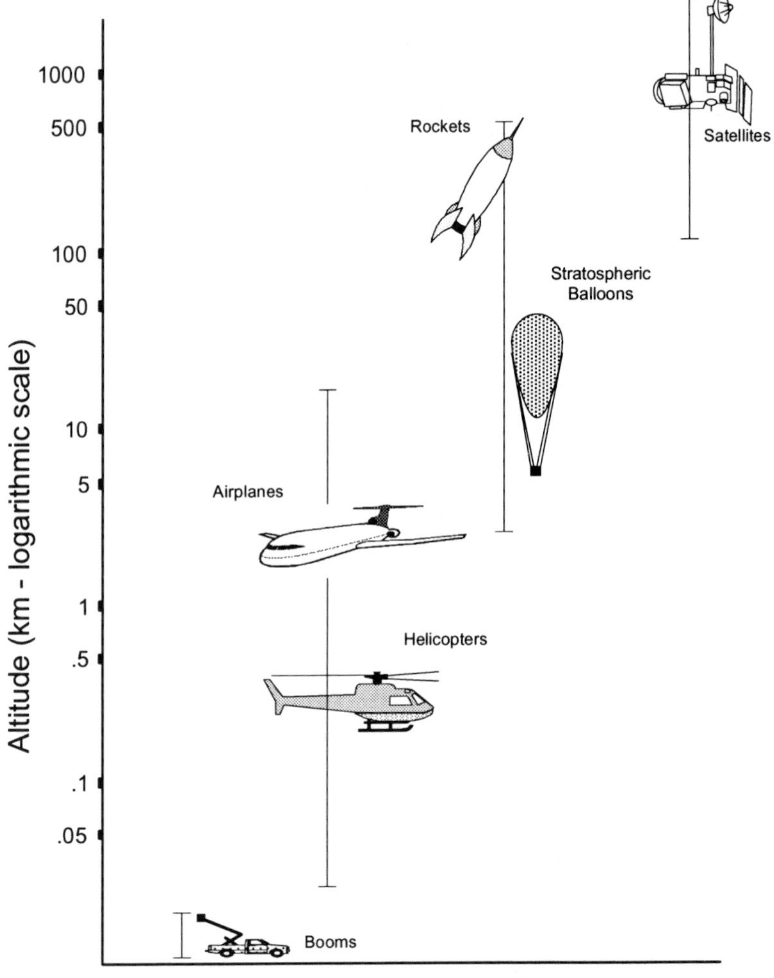

Figure 4.1 Height above sea level of typical remote sensing platforms.

Soyuz space station and earth resource monitoring and meteorological satellites (e.g., Landsat, SPOT, NOAA AVHRR).

Geosynchronous orbiting or geostationary platforms are placed in very high orbits (36,000km) over the equator. At this altitude platforms orbit in synchrony with the earth's rotation, and thus remain fixed over the same location rather than sweeping over the landscape as do platforms in lower orbits. An international network of five geostationary satellites positioned in orbit around the equator at 72° intervals have provided global meteorological information since 1976. The United States owns two: the Synchronous Meteorological Satellite (SMS, 140°W) and the Geostationary Operational Environmental Satellite (GOES, 70°W). Meteosat (0°) is owned by the European Space Agency. The USSR (70°E) is owned by Russia. The Geosynchronous Meteorological Satellite (GMS, 140°E) is owned by Japan.

The primary sources of remote sensing information are based on three major technologies: EMR-sensitive chemicals (photography), EMR-sensitive electronics (panchromatic and multispectral imagery), and radar. The resolution and spatial coverage of remote sensing systems varies with technology and the platform used (table 4.1). At present, aerial photography provides the highest resolution source of imagery, whereas satellite remote sensing provides the greatest coverage of global remote sensing data. Low-orbiting platforms for earth resource monitoring and meteorology circle the globe approximately from pole to pole. As the earth rotates slowly eastward, each perpendicular orbit of the platform will provide the remote sensing system with a different view of the earth's landscape—a view located to the west of the previous orbit. Because the angle of view (width of terrain visible) of sensors on board polar orbiting satellites (platforms) is fixed, and because orbital paths are spaced according to the circumference of the earth at the equator, areas located at mid- to high-latitudes will fall within the field-of-view of the sensor more frequently than those at lower latitudes.

When selecting between low altitude (terrestrial and airborne) sensor systems and satellite imagery, one of the first issues to address is scale. Terrestrial and airborne sensors offer very high spatial resolution but at a cost of providing only very localized simultaneous coverage. Satellite sensors have a much more synoptic (regional or continental) view of the landscape but can only do so at a relatively coarse spatial resolution. Because of the trade-off between spatial resolution and area coverage (table

Table 4.1
Image Coverage and Resolution Trade-Offs

Source of Imagery	Resolution	Coverage (width)
Aerial Photography (240 mm format film)		
1:10,000	<0.2m	2km
1:50,000	3m	10km
1:100,000	5m	15km
1:150,000	10m	25km
1:250,000	15m	35km
Satellite Photography		
Soyuzkarta MK-4	6m	120–270km
Satellite Imagery		
SPOT Panchromatic	10m	60km
SPOT HRV	20m	60km
ALMAZ-SARadar	15–30m	40km
LANDSAT TM	30m	185km
LANDSAT MSS	80m	185km
AVHRR	1km	3,000km
METEOSAT	2.4km	>10,000km

4.1), selection of an appropriate sensor system will depend on whether the question to be addressed is highly localized and refined, or more regionalized and coarse grained. In addition to the spatial scale trade-off, we also need to take into account several distinct advantages of satellite remote sensing over airborne or terrestrial platforms. Satellite remote sensing systems offer

- worldwide coverage without political or security restrictions in most countries;
- frequent and repeated coverage of the same area;
- consistency in the manner that radiation reflectance and emission data are recorded; and
- per unit area costs of obtaining and interpreting satellite imagery at least an order of magnitude less costly for areas larger than 1,000km^2.

Analog (Photographic) Systems

The sensors on terrestrial, airborne, orbiting, and geostationary platforms record the radiation reflected or emitted from objects at or near the earth's surface in either analog (continuous) or digital (stepwise) form.

Table 4.2 provides a comparison of various aspects of analog versus digital remote sensing.

The most widely available analog remote sensing imagery is photography. Photographic film is composed of a single layer of silver halide emulsion (black-and-white or panchromatic) or triple layers of yellow, magenta, and cyan emulsions (color). When exposed, molecules within the film's emulsions react to the radiation emanating from the landscape, such that once developed the film produces a black and white or color representation of the continuous range of light intensity evident across the scene. Panchromatic and color films are sensitive primarily to light in the visible range. Films developed specifically for remote sensing analysis are sensitive to reflected-IR light, and are used extensively for vegetation mapping and analyses.

Photographic remote sensing is primarily obtained by commissioning specific aerial overflights of areas of interest, or by locating prints within archives of aerial photography obtained systematically for national cartographic mapping purposes. The USGS 1:58,000 color IR National High

Table 4.2
Analog Versus Digital Camera Imaging

	Photographic Film	Electronic Image
Sensor	Silver halide film	Photodiode array (CCD)
Spectral range	Visible to near IR	Visible to near IR
Resolution	~2.5–3.0 million pixels (35mm)	400,000–1.5 million pixels
Shutter speed	≤1/8000 second	≤1/8000 second
Light sensitivity	≤1600ASA	≤400–1600ASA
Image generation	Chemical developing/printing	Computer hardware/ software
Storage medium	Transparency or Print film	Magnetic, optical, solid state
Analysis	Visual	Digital and visual
Image transmission	Mail, FAX	Modem, computer networks
Display	Projected slides	Computer monitors, projection TV
Hardcopy	Silver halide prints	Color ink jet, thermal wax, electrostatic
Camera price	$200–$3000	$900–$9000
Media cost	$15 per 36 exposures	$0–$5

Altitude Photography Program is an example of a nationwide archive of high resolution aerial photography.

A large-format camera (LFC) was developed by NASA specifically for the manned space shuttle. The camera uses 23×46cm panchromatic, true color, and IR film, and has a spatial resolution of approximately 5–20m depending on the altitude of the shuttle. To date, only 2,160 photographs have been obtained, and the shuttle does not carry the camera routinely. The only operational orbiting photographic system is the Russian Soyuzkarta program. The KFA-1000 camera has a spatial resolution of 4–7m and uses a two-color emulsion film sensitive to 560–670 and 670–810nm wavebands. The KATE-200 is a 15–30m spatial resolution, multispectral camera. Optical filters transmit terrain brightness only within three wavebands (500–600, 600–700, and 700–800 nm). The landscape brightness for each waveband (*channel*) is then recorded on a separate panchromatic negative. The MK-4 is the most sophisticated of the Soyuzkarta large format cameras. It has a resolution of approximately 6m and records terrain brightness on three separate panchromatic negatives chosen from a suite of five spectral channels (400–700, 460–505, 515–565, 580–800, and 635–900 nm). All Soyuzkarta photography has a 40–60% longitudinal overlap, allowing for viewing in stereo and the development of digital terrain maps. Soyuzkarta photography is distributed through SPOT Image Corporation and Hughes STX.

Photographic imagery has traditionally been interpreted visually. The boundaries of features detected are drawn by hand or, more recently, using a cursor and digitizing tablet (figure 4.2). Within the last ten years, advances in scanning technology have made it much more practical to convert whole analog photographs into digital images (MicroImages 1991). The most common scanners are rotating or flatbed densiometers capable of 1–12μm resolution, and the more recent and much less expensive linear photodiode or charge-coupled-device flatbed scanners that have a resolution of 4–12μm (DBA Systems Inc., Melbourne, Florida, tel: 407-727-0660). If high resolution output is not essential, then a video frame grabber can capture images from a video camera at resolutions of up to 1,024×768 pixels in 24 bit color or 256 (8 bit) graytones.

Scanning and digital processing of photography, unlike visual interpretation, makes full use of the spectral detail contained within the photograph, and takes advantage of the feature enhancement and extraction facilities available within most modern image processing software pack-

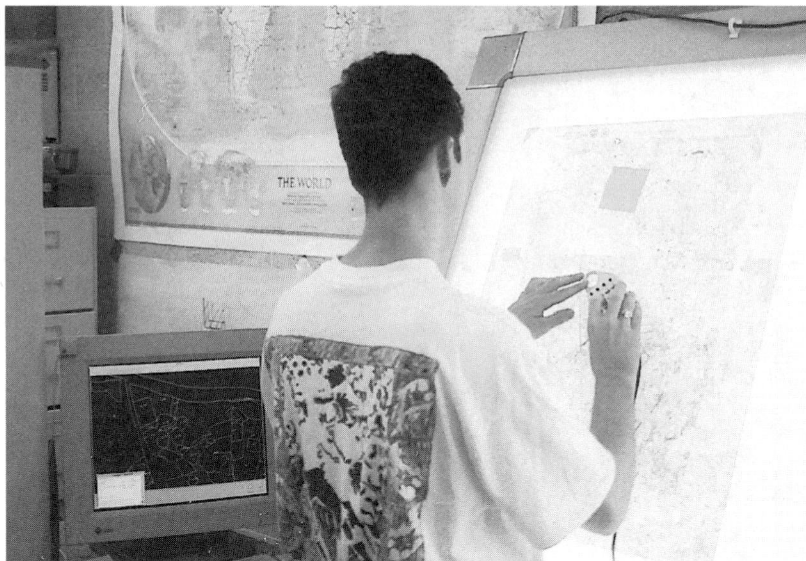

Figure 4.2 Manual digitizing of map or image information. The operator is using the digitizer puck (a device with input buttons and a crosshair cursor) to trace the outlines of features of interest. The digitizer records the movement of the puck and transfers the digital outline to the computer (*lower left*) that stores and displays the data.

ages. Scanning cannot, of course, increase the spatial or spectral resolution of the original photography.

Digital Systems: Scanning Sensors

Digital sensors record radiation reflected or emitted from a landscape, either by systematically scanning across the landscape and progressively building an image (*scene*) as a grid of discrete radiation brightness values recorded for small rectangular landscape blocks (figure 4.3) or by recording landscape brightness over a large area as a snapshot, one frame (image) at a time. Most digital imagery is obtained from scanning sensor systems. In these systems a detector records the brightness (related to the number of photons striking the detector) of the small landscape block within its instantaneous field-of-view (IFOV), the size of which is determined by the altitude of the satellite, the optics that focus the radiation

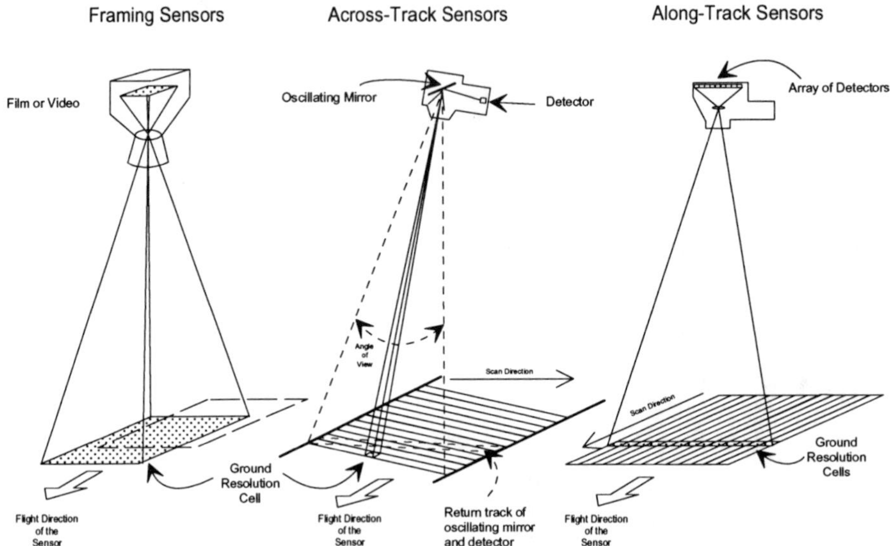

Figure 4.3 Terrain imaging using scanning and framing sensor systems. The framing sensor system (*left*) records the brightness of the terrain as an instantaneous snapshot (i.e., a photograph or as one frame on a videotape). In contrast, the across-track and along-track sensors build up the image sequentially as a series of lines of pixels. The along-track sensor uses an array of detectors equal to the number of pixels in a line of data to record the brightness of terrain in a whole line of data at one time. The across-track sensor uses an oscillating mirror to scan a line of terrain brightness data one pixel at a time.

on the detector, and the physical size of the detector itself. The scanning system is able to record the brightness of the whole landscape by sweeping the detector rapidly across the terrain.

The brightness signal (voltage) recorded by the detector varies continuously as the sensor's optics and mirrors sweep the IFOV over the landscape. This analog signal is amplified and then converted to digital form by sampling the signal strength of the detector at standard intervals. In this way an image is built up from a series of adjacent cells or picture elements (pixels), the size of which is a function of the IFOV of the detector and the rate at which the analog detector signal is digitized. If, for example, the sampling rate is faster than the motion of the scanner over the landscape, then the IFOV of the detector at time t+1 will overlap a portion of the terrain sensed during time t. In this situation the pixel size will be smaller than the IFOV; however, the brightness of each pixel will still be based on the radiation from the terrain within the IFOV of the detector (figure 4.4).

Digital images are either panchromatic or multispectral. A panchromatic, or single band, digital image is composed of a rectangular mosaic of pixels. The value of each pixel represents the brightness of a discrete area of the landscape within a waveband determined by the sensitivity of the detector. The waveband sensitivity of the detector can be broad (including, for example, all wavelengths of visible light from violet to red) or very narrow (wavelengths, for example, only in the visible green spectrum between 0.45 and 0.50μm). A panchromatic image, depending on the detector used to obtain the data, can duplicate the information found within a black-and-white aerial photograph, or it can represent only the blueness, or redness, or reflected IRness of the landscape. Multispectral imagery is very simply composed of a set of exactly overlapping panchromatic images each of which represents a different spectral waveband.

We will describe four types of digital image scanning systems: (1) across-track scanners, which use pivoting mirrors to record landscape brightness as individual pixels connected into lines running perpendicular to (across) the orbital track of the platform; (2) spin scanners, which record landscape brightness as lines of pixels by spinning the sensor platform on its axis; (3) along-track scanners, which record landscape brightness simultaneously along a whole line of pixels arranged sequentially along the orbital track of the platform; and (4) oblique scanners, which transmit their own energy source to one side of the flight path of the platform, generating an image by recording when and how much energy is reflected back from the landscape. Last, we will describe another type of digital remote sensing technology that uses framing sensors. The various kinds of framing sensor are distinguished from all four kinds of scanning system in that the brightness of the landscape is recorded instantaneously—it is not built up through a progressive series of scans.

Across-Track Scanners

The first earth resource monitoring satellites employed a west-to-east across-track scanning system. The Landsat MSS and TM systems are both examples of across-track scanners. Within the Landsat MSS scanner are

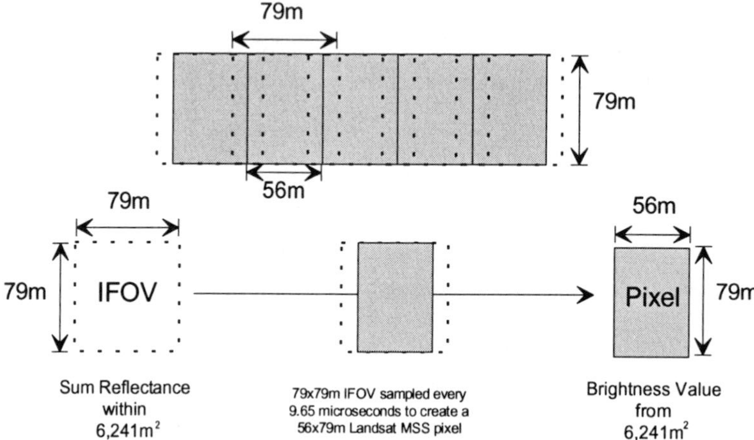

Figure 4.4 Difference between the IFOV and pixel size for Landsat MSS imagery. As each Landsat MSS sensor is square, the satellite's altitude and optics yield an instantaneous-field-of-view of 79×79m. As the scan mirror oscillates from west to east during the satellite's southward orbit, the voltage that represents the radiation brightness striking the sensor is sampled every 9.95×10^{-6} seconds, resulting in 3,300 digital voltage values being recorded along the 185km scan line. Thus although the MSS IFOV is 79×79m, the sampling rate effectively creates an MSS pixel (picture element) of 56×79m. This means that while the MSS pixel is 4,424m^2, the digital reflectance value associated with it comes from an area of 6,241m^2.

six radiation detectors for each of the four spectral bands that comprise an MSS image—a total of twenty-four detectors. As the satellite travels north to south in its near-polar orbit, an oscillating mirror focuses radiation on the detectors as though they were sweeping across the landscape in a 185km-wide swath from west to east. The scanner's optics focus radiation reflected from a 79×79m area on the ground onto the end of a 24-strand fiber optic cable that channels the radiation to each of the detectors. The 79×79m instantaneous field of view (IFOV) of the Landsat MSS sensor is determined by the altitude of the satellite (platform) and the optics that focus radiation on the detectors. Each detector registers the intensity (brightness) of the radiation within its waveband sensitivity as a voltage that varies continuously (i.e., as an analog signal) as the detector is swept across the landscape. This analog voltage is sampled (digitized) every 9.56 microseconds and converted to an integer value, which is either recorded on tape onboard the satellite or transmitted immediately to a receiving station on earth. This digitizing process generates a series of picture elements (pixels) than form a line of discrete radiation brightness values along the scan direction from west to east. With six detectors for each spectral band, each west-to-east scan generates six lines of brightness information. The Landsat MSS sensor does not record information on the east-to-west return of the oscillating mirror because as the satellite is orbiting from north to south, the swath width and mirror oscillation rate are set so that scanline 1 abuts scanline 6 from the previous scan (figures 4.3 and 4.5).

The major disadvantage of across-track scanners is the relationship among IFOV—swath width—and the time that the detector is exposed to the radiation from any given area of terrain (dwell time). The smaller the IFOV and the wider the swath width, the shorter is the dwell time. The shorter the dwell time, the smaller is the radiant flux (energy) that strikes the sensor (table 4.3). If dwell time over a given pixel is too brief, the brightness signal recorded by the sensor may not be recognizable from the background electronic noise associated with any electronic component. As a result, spatial resolution and area coverage of across-tracking imagery is constrained by detector sensitivity and dwell times. Examples of polar orbiting across-track scanners are the NOAA Advanced Very High Resolution Radiometer (AVHRR), the discontinued Nimbus 7 Coastal Zone Color Scanner (CZCS), and the Landsat Multispectral Scanner (MSS) and Thematic Mapper (TM). Airborne scanners are primarily experimental and include the Advanced Visible Infrared Imaging Spectrometer (AVIRIS), the Thermal Infrared Multispectral Scanner (TIMS), and the Calibrated Airborne Multispectral Scanner (CAMS).

Spin Scanners

Geostationary meteorological satellites, such as GMS (Japan), Synchronous Meteorological Satellite (SMS-USA), Geostationary Operational Environmental Satellite (GOES-USA), and METEOSAT, remain perpetually above the same spot on earth. As the earth does not move below them, each platform spins on a north-south axis to scan the terrain below. With each rotation of the satellite, the sensor records a single line of brightness data running parallel to the equator. By changing the angle of

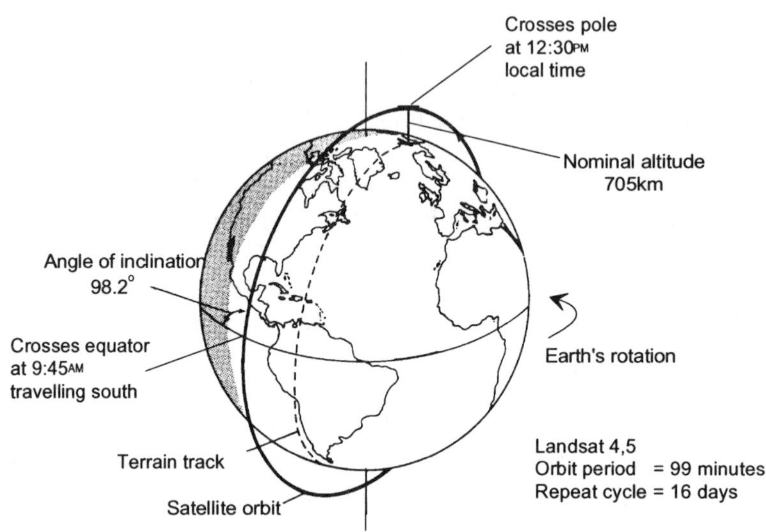
Crosses pole
at 12:30PM
local time
Nominal altitude
705km
Angle of inclination
98.2°
Crosses equator
at 9:45AM
travelling south
Earth's rotation
Terrain track
Satellite orbit
Landsat 4,5
Orbit period = 99 minutes
Repeat cycle = 16 days

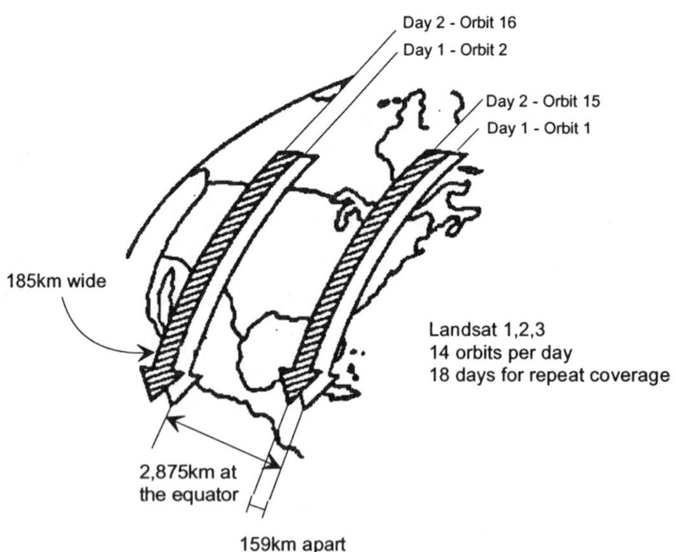
Day 2 - Orbit 16
Day 1 - Orbit 2
Day 2 - Orbit 15
Day 1 - Orbit 1
185km wide
Landsat 1,2,3
14 orbits per day
18 days for repeat coverage
2,875km at
the equator
159km apart

view with a mirror and stepping motor, adjacent lines of data are collected with each rotation, building up a full hemispheric image every thirty minutes. Visible infrared spin-scan radiometer (VISSR) is the name of the sensor carried by the GOES satellites.

Along-Track Scanners

The Satellite Pour l'Observation de la Terre (SPOT) provides a commercial source of along-track scanner imagery. Along-track scanners are often called "pushbroom" systems because the array of the detectors that simultaneously record landscape brightness along a line of pixels are effectively pushed like a broom along the orbital path of the platform. The SPOT satellite was the first operational along-track scanning system that took advantage of the new *charge coupled device* (CCD) detectors that can be installed in a linear array equivalent to the swath width of the scanner (figure 4.3). Charge coupled devices are solid-state chips that generate a voltage proportional to the amount of EMR that strike them. Arrays of CCD detectors record the brightness of a series of pixels simultaneously along a scan line. This technological advance eliminated the need for an oscillating mirror and thus increased the dwell time possible over each pixel.

The greater effective dwell time of pushbroom scanners increases the detector's signal-to-noise ratio, which allows for both a smaller IFOV, with concomitantly finer spatial resolution, and use of detectors with much narrower radiation bandwidths, thus increasing the spectral resolu-

---

Figure 4.5 The near-polar orbital pattern of the Landsat satellites. All polar orbiting satellites circle the earth multiple times each day; they use the earth's west to east rotation to image a different swath of the terrain with each orbit. Landsats 1, 2, and 3 completed 14 orbits per day. Each orbit placed the satellite 2,875 km west of its previous orbit (at the equator). The orbiting rate (103 minutes), and image width (185km) allowed each area on earth to be imaged every 18 days. As the circumference of the earth reduces toward the poles, overlap between images is greater at higher latitudes and thus revisit frequency is higher. The Landsat satellites are sun synchronous—meaning that the local time that they image the same point on earth remains constant. Landsats 4 and 5 cross the equator at 9:45 A.M. local time. Landsat 1, 2, and 3 crossed the equator at 8:50 A.M., 9:08 A.M., and 9:31 A.M. local time, respectively.

Table 4.3
Factors Affecting Sensor Signal Strength

Radiant flux	The quantity of energy reflected or emitted from the terrain per unit time is the radiant flux. This value can change with season and atmospheric conditions.
Altitude	Radiation flux (signal strength) reaching a detector is inversely proportional to the square of the distance from the feature to the detector. Signal strength and altitude are inversely related.
Bandwidth	Signal strength declines as the waveband sensitivity of a detector narrows, because signal strength is related to the sum of the radiation flux over all wavelengths that strike the detector.
IFOV	The size of the detector and the optics of the scanner determine the IFOV. The IFOV and signal strength are positively correlated.
Dwell time	The time spent detecting the brightness of a given area of terrain is the dwell time. Dwell time and signal strength are positively correlated.

SOURCE: Adapted from Sabins 1986:16.

tion of the imagery. Additional advantages of detector array systems are lower power requirements, no moving parts, greater life expectancy, higher geometric and radiometric accuracy, and lower overall cost of development and operation. Pushbroom systems have, however, two disadvantages: (1) a large number of detectors must be calibrated relative to one another to avoid systematic detector bias within the imagery and (2) CCD detectors have not been developed that can sense wavelengths longer than near-IR. Two additional pushbroom sensor systems, the Japanese Marine Observation Satellite (MOS-1) and the Indian IRS-1 are now in orbit, and both systems are currently sending data to several receiving stations worldwide.

Oblique Scanners (Radar)

Oblique scanning systems are primarily active systems that transmit microwave energy (radar) or sometimes sound waves (sonar) at an angle to the platform, and record when and how much energy is reflected back from the terrain. However, the SPOT along-track system is also pointable, enabling it to image sections of terrain at an oblique angle to the along-track direction. By imaging the same section of terrain from both sides (east and west), a stereoscopic image can be constructed from which a digital elevation map can be generated. Radar (radio detection and ranging) is far more commonly used than sonar (sound navigation and ranging), and so radar scanners are the topic of this section.

Radar systems transmit microwave radiation in the 1–30cm range. Because radar is largely unaffected by clouds, and because it actively "illuminates" the landscape, radar systems can operate at all times during the day and night and in all weather conditions. Radar imagery therefore has great potential for, among other things, natural resource inventories and monitoring in habitually cloud-covered tropical areas for which multispectral imagery can be exceedingly difficult to obtain.

A radar sensing system works by transmitting a pulse of microwave radiation, then recording both the strength of the pulse reflected back from the landscape (detection) and the time for the reflected pulse to return (ranging). The strength of the returned pulse depends on the form and composition of the landscape object (the scatterer) that scatters and reflects the microwave pulse. The time to receive the return signal depends on the object's distance from the radar system. The resolution of radar systems is determined by the beam width of the microwave pulse generated by the system. Beam width (resolution) is inversely proportional to the range of the scatterer and the wavelength of the transmitted microwaves; it is directly proportional to the size of the antenna. For example, to resolve two objects 30m apart and located 5km from the radar system, a 3cm microwave radar would require an antenna with an aperture width of >0.5m. A 20cm microwave radar would require an antenna 3.3m wide to distinguish the two objects. Although shorter microwaves (<0.5cm) would provide higher resolutions, they lack the weather penetration capacity that is the major advantage of active remote sensing systems.

The resolution of radar imagery is determined by this equation:

$$Resolution\ (m) = \frac{range(km) \times wavelength(cm)}{antenna\ aperture(m)}$$

Meteorologists are able to use large rotating antennae installed on the ground to create the circular images of rainfall familiar to watchers of television weather reports. Mounting a large rotating antenna on an aircraft is clearly impractical. However, the same resolution as a large rotating antenna can be achieved by mounting a fixed antenna of equal length that points only to the side of the aircraft. Side-looking airborne radar (SLAR) is the most commonly used technique for obtaining radar images of the earth's terrain.

In SLAR systems a pulse of microwaves is transmitted to the side (across track) of the aircraft's line of flight (along track) at a particular oblique angle to the terrain (incidence angle). Time delay in reception of the reflected signal provides the slant angle distance (range) of the scatterer (object). The intensity of the return signal contains information about the scattering characteristics of the object. The forward movement of the aircraft is synchronized with the pulses of microwaves to illuminate the next strip of terrain, and thus compile a complete radar image of the land surface. Returning signal intensities from each microwave pulse control the brightness of a spot recorded on photographic film or the voltage recorded on magnetic tape. The position of the spot along the scan line corresponds to the return time of the signal, and indicates the range of the object. Movement of the film or magnetic tape is synchronized with the along-track velocity of the aircraft (figure 4.6). Inertial navigation systems maintain SLAR aircraft at a constant altitude and ensures that each path line is almost precisely parallel to the preceding one. Inertial navigation systems are also able to control the attitude of the antenna to counter the yaw, roll, and pitch of the aircraft, thus maintaining a constant incidence angle.

Resolution in the across-track direction is determined by the pulse duration and the incidence angle. Spatial resolution is equal to one-half the

---

Figure 4.6 Radar imaging of landscapes using SLAR. The SLAR (side-looking airborne radar) form of imaging uses real aperture radar to generate an image of the landscape to the side of the flight path of the aircraft. An image is created based on the time and strength of the radar signal that is returned from objects in the landscape. The return time of the radar signal represents the distance of the terrain feature from the radar antenna (sensor) mounted on the aircraft, and determines the position of the feature on the image. The strength of the return signal is determined by the roughness and orientation of the feature—buildings and metal bridges reflect strongly, forests are diffuse reflectors, and open water is a specular (mirrorlike) reflector from which little if any of the signal returns to the antenna. The stronger the return signal, the brighter is the point on the final image.

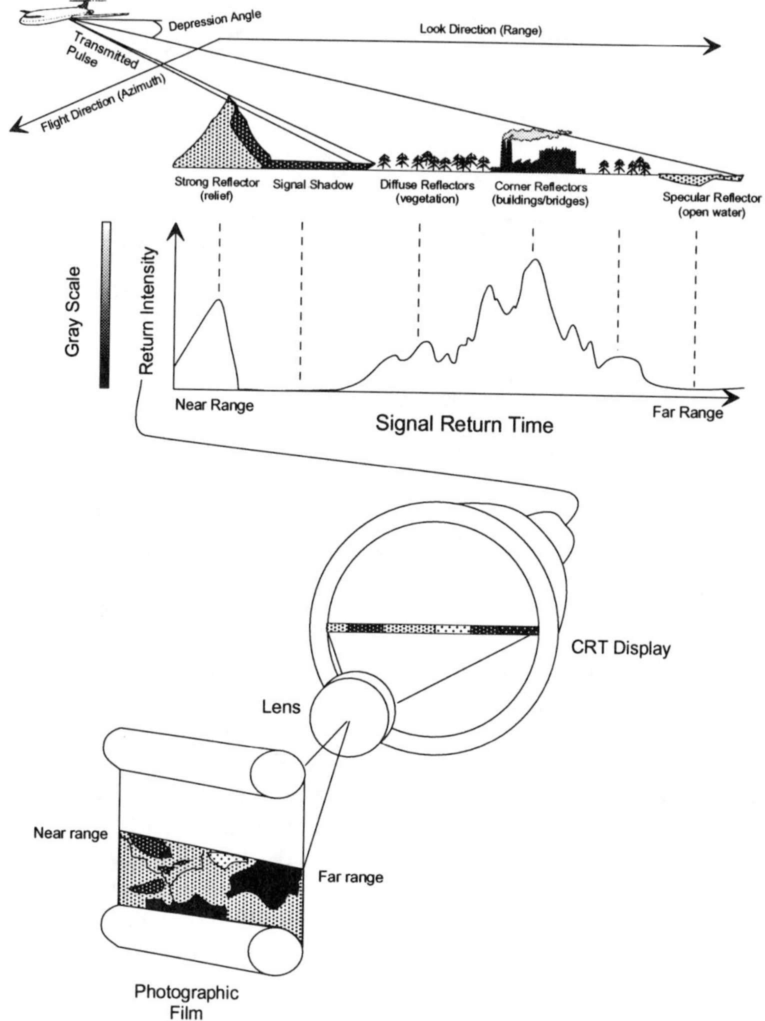

pulse length, which is calculated by multiplying the pulse duration by the speed of light. For example, a 10^{-7} second pulse is 30 meters long, resulting in a theoretical spatial resolution of 15 meters. The pulse duration resolution is modified by the incidence angle of the transmission. As figure 4.7 shows, resolution increases as the incidence angle becomes greater, that is, as the radar image becomes more oblique. Objects that are closer

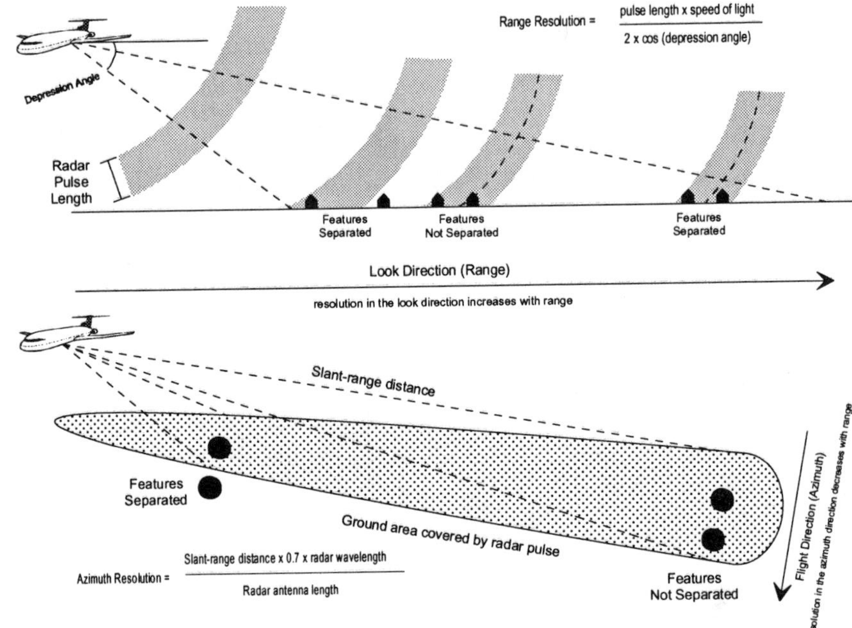

Figure 4.7 Resolution of SLAR (side-looking airborne radar) sensors in relation to range and azimuth. *Top:* Two features will be seen as discrete objects in the look direction (perpendicular to the flight path of the aircraft) if they are separated by a distance at least half the length of the radar pulse. Note how the left-most pair of buildings will be detected as two objects, whereas the middle pair will be detected as one large object. Radar resolution in the look direction is greater when the depression angle is small. As the depression angle decreases with the range, the right-most pair of buildings can be detected as two separate objects, even while the middle pair cannot. *Bottom:* Real aperture radar resolution in the azimuth direction (parallel to the flight path) decreases with range. As a result, though both pairs of objects have the same internal separations, only the left-most pair is detected as two discrete objects.

together than approximately half the pulse length will be perceived by the antenna as one broad scatterer, rather than as two discrete objects.

The ability of SLAR to distinguish two objects separated from one another in the along-track direction of the aircraft (but in the same across-track range) is determined by the beam width of the system. If two objects fall within the beam width of the SLAR system, they are recorded as one scatterer. Thus the wider the beam width, the further apart objects have to be for them to be detected as separate objects. Beam width becomes narrower as the length of the antenna increases. Beam width fans out (increases) as the radar signal travels further away from the aircraft. As a result, a 3cm microwave radar with a 500cm antenna has an along-track resolution of 60m at 10km, and 180m at 30km. As beam width of SLAR systems increases with range, the width of the strip of landscape that can be imaged by a SLAR system is determined primarily by the length of the antenna and the resolution desired; the longer the antenna and/or the coarser the resolution, the wider the area that can be imaged at one time. Although this is not a severe constraint to radar imaging from aircraft, it does preclude using SLAR techniques on satellites, as an orbit of 900km would require an antenna 135m long to obtain 20m resolution.

The advent of synthetic aperture radar (SAR) has overcome the antenna length problem associated with real aperture SLAR. The SAR system is distinguished from the SLAR system in that it receives return signals from ahead, perpendicular to, and behind the aircraft, and takes advantage of the motion of the radar platform to make a short antenna with a wide beam width act like a very long antenna with a proportionally narrower beam width. As the radar platform moves in the along-track direction, it transmits a series of wide beam pulses at regular intervals. As the platform passes an object on the terrain below, the feature will enter the microwave beam and scatter back radiation for as long as it remains in the beam. As the beam width increases with distance from the transmitter, distant objects on the terrain will remain within the microwave beam and reflect radiation back to the antenna for longer durations than do objects close to the transmitter.

Because the more distant objects are detected for longer durations than are objects close to the antenna, the effective length of the SAR antenna appears longer for distant objects. In fact, SAR antenna length is proportional to the object's range. And because spatial resolution of radar is proportional to the length of the antenna and inversely proportional to the range, SAR technology compensates for the two opposing effects, and the resolution in the along-track direction remains the same at all ranges. The SAR form of radar imaging is particularly useful because it allows for high resolution radar imagery even from satellite altitudes. Like SLAR, across-track resolution of SAR imagery is determined by pulse duration and incidence angle.

Imagery from SAR is recorded as a hologram composed of diffraction patterns. Much as a police siren changes pitch as it approaches, passes, and recedes, the wavelength of the signal returning from a scatterer will vary with the movement of the aircraft—becoming shorter on approach and longer on retreat. If the returning signal of variable wavelengths detected by the antenna is then combined with a coherent (single wavelength) reference signal, an interference pattern is created. When the returning pulse is in phase with the reference signal, the interference is constructive and the recorded voltage is high. When out of phase, the interference is destructive and the resulting voltage will be low and the corresponding spot on the photographic film will be dark. Interference is recorded not as a circular pattern of dark and light lines, but as a narrow strip or cross-section (figure 4.8).

The width of the strip corresponds to the pulse length. As with SLAR, the range of the object in the across-track direction is recorded as the distance of the interference cross-section from the near edge of the film or magnetic tape. When the SAR hologram is illuminated with coherent light (using a cylindrical lens to focus the across-track information), a detailed image of the terrain is generated.

Although all radar images are oblique, displacement of elevated areas toward the antenna gives the impression that the image was viewed from directly overhead. This displacement (image layover, or lean) occurs because radiation reflected from elevated areas returns to the antenna sooner than does radiation reflected from regions lower on the elevated object or even somewhat in front of it (figure 4.9). This effect is analogous to objects coinciding on an oblique aerial photograph when they exhibit the same angular coordinates (i.e., two objects 50m apart but with one directly behind the other).

Elevated features also create signal shadows in the across-track direction by blocking microwaves. If the far-side slope angle is at, or greater than, the radar view angle, little or no backscatter will occur; the area will appear as a radar shadow (figure 4.6). The length of the shadow increases with the elevation and range of the object. In some cases image layover and shadow effects combine to completely hide terrain features. Shadows can be reduced by decreasing the incidence angle. This, however, increases the distortion caused by image layover. Because incidence angle, range, and object elevation all affect shadowing and image layover, it is not possible to eliminate these two forms of distortion. Compensating for

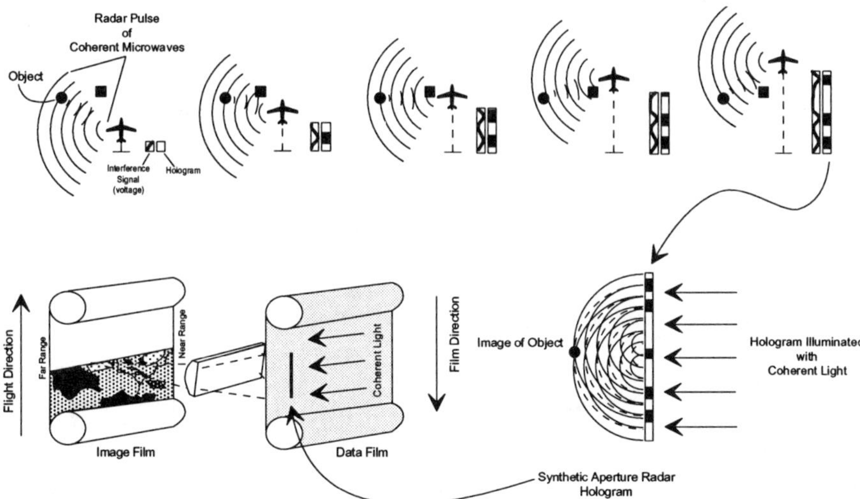

Figure 4.8 Holographic imaging in SAR airborne radar systems. Synthetic aperture radar is different from real aperture radar used in SLAR systems in two ways. First, the return signal data are stored on film as a hologram that, when illuminated with coherent light, generates a visually interpretable image of the landscape. Second, the spatial resolution of SAR is constant in the azimuth direction regardless of the range of the objects. Notice how the square object closer to the aircraft does not remain within the radar beam for as long as the more distant object.

one exacerbates the problems caused by another. Nevertheless, if the terrain has little relief, increasing the incidence angle will minimize layover without resulting in excessive shadowing.

Digital Systems: Framing Sensors

Framing systems are distinguished from across-track scanners, spin scanners, along-track scanners, and oblique (radar) scanners in that they record landscape brightness as an almost instantaneous snapshot rather than by building the image from a series of separate scanlines. Three major kinds of framing sensors have been used to gather remote sensing imagery: satellite return beam vidicon, aerial videography, and aerial digital photography.

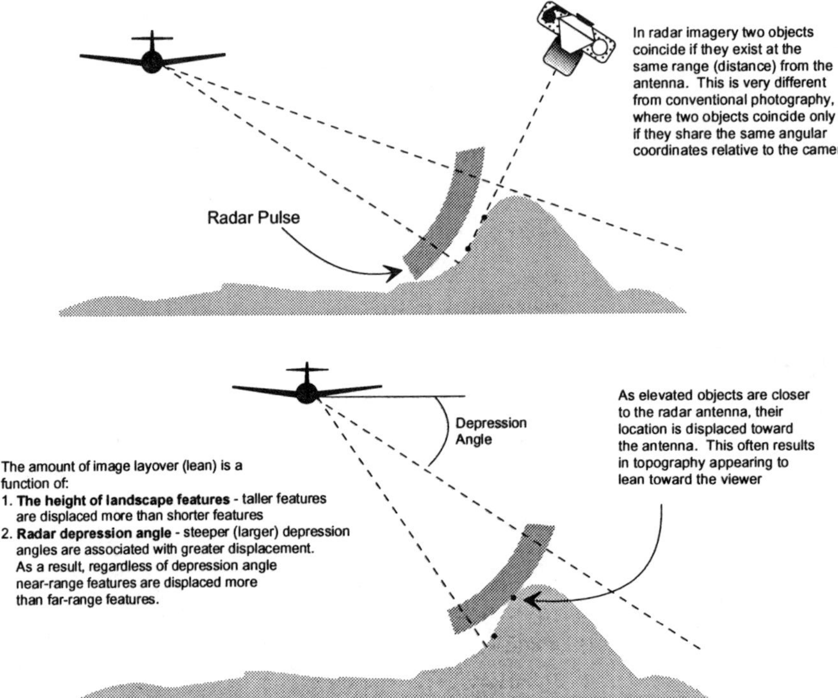

In radar imagery two objects coincide if they exist at the same range (distance) from the antenna. This is very different from conventional photography, where two objects coincide only if they share the same angular coordinates relative to the camera.
Radar Pulse
Depression Angle
As elevated objects are closer to the radar antenna, their location is displaced toward the antenna. This often results in topography appearing to lean toward the viewer
The amount of image layover (lean) is a function of:
1. The height of landscape features - taller features are displaced more than shorter features
2. Radar depression angle - steeper (larger) depression angles are associated with greater displacement.
As a result, regardless of depression angle near-range features are displaced more than far-range features.

Figure 4.9 Layover, or "lean," of elevated objects when imaged by radar systems. All radar systems image the landscape at an oblique angle. However, because the return signal of elevated objects arrives before that of objects on lower slopes, topography appears to lean toward the viewer, giving the impression that the image was taken from directly overhead.

Satellite Return Beam Vidicon

The first major digital framing sensor developed for natural resource monitoring was the Return Beam Vidicon (RBV), which was flown on the first three Landsat satellites. A vidicon is ostensibly a television tube camera that records successive overlapping scenes by closing and opening a shutter, thus allowing light to fall on the charged surface only at selected intervals. The phosphors on the charged surface of the RBV sensor are sensitive to radiation wavelengths (0.5–0.85 μm) that approximate IR photographic film, and the resulting data can be displayed as a panchromatic image.

Landsat 1 and 2 employed three RBV sensors, each with a special filter that restricted waveband sensitivity (blue-green, yellow-red, red-IR), and allowed generation of composite multispectral images. Landsat 3 employed a two RBV panchromatic sensor system. Data from the RBV sensors are only available from June 23, 1972 to September 7, 1983 when Landsat 3 was retired. One major use of Landsat 3 RBV imagery was to enhance the visual appeal of a Landsat MSS image. The higher resolution RBV imagery (24m), when merged with Landsat MSS (79m resolution) maps, conveys the appearance of more information because, as we have noted before, more detail does not necessarily mean more information.

Aerial Videography

The second kind of digital framing sensor, aerial videography, differs from RBV because videoimages can be viewed easily both as single frames and as continuous moving images, using equipment available in any electronics store. Aerial videography is becoming progressively more important as a quick, relatively inexpensive tool for gathering color or panchromatic information at various scales over relatively large areas.

In 1988 the American Society of Photogrammetry and Remote Sensing held its first workshop on remote sensing applications of videography. The U.S. Fish and Wildlife Service, the U.S. Department of Agriculture, the U.S. Forest Service and the (newly created) National Biological Service are now all using aerial videography to provide inexpensive data for natural resource management. Aerial videography has already been used to help answer a variety of questions. Sidle and Ziewitz (1990) used a color video camera mounted on a Cessna 172 aircraft to map and monitor piping plover nesting habitat along 400km of the Platte River. Everitt et al. (1991) used color infrared (CIR) video to delineate black mangrove communities on the Gulf Coast of Texas. Another research group used multispectral video to generate a land use geographic information system (GIS) for a region of Senegal (Marsh et al. 1990) and to verify a Landsat landcover classification in the Brazilian Amazon (Marsh et al. 1994).

Pre-1980s video cameras, like the RBV sensor, record the brightness of an area of terrain on the charged phosphor-coated surface of a vidicon tube. Color cameras use three vidicon tubes that with the aid of three filters record red, green, and blue brightness levels. An electron beam sweeps the phosphor surface of each television tube to record, as a voltage, the charge differences that represent the radiation brightness differences evident in the landscape within the camera's field of view.

The analog signal generated by the vidicon is specified by industry standard. In the United States and Japan this is NTSC, in France and eastern Europe SECAM, and in the United Kingdom and the rest of Europe PAL. The NTSC RS-170 standard was designed to avoid interference with the 60Hz AC household electrical current that is standard in the United States. As a result, NTSC vidicon cameras scan alternate lines of an image frame every 1/60 second, generating a full interlaced image every 1/30 second. The NTSC standard scans 525 lines per frame (262.5 lines per field), twenty of which are not used in each field, in order to allow time for the electron beam to reposition from the bottom to the top of the image.

The slow "shutter speed" (1/30 second) of vidicon cameras means that, even at the relatively slow ground speed of small single-engine aircraft, image motion will occur between scanning of the two fields that comprise a full frame. This would result in unacceptable blurring of the image. To minimize blurring in an NTSC system, often only 242 of the 485 available lines are used for analysis. The PAL and SECAM systems have higher resolution (625 lines per frame) corresponding to the longer frame acquisition rate associated with the European 50Hz AC current standard.

Though vidicon cameras can generate an analog signal with 485 lines per frame, the carrier frequency of VHS video recorders (4.4MHz) limits playback to 240 lines per frame. The more recent Super-VHS and Hi-8 formats, with a carrier frequency of 7.0MHz, can record 430 lines per frame. Furthermore, in order to maintain compatibility with black-and-white television signals, video recorders cannot record simultaneous red, green, and blue (RGB) signals, and must superimpose the three signals into a composite color signal. The frequency used to encode the composite color signal limits recorded resolution to 250–320 lines per frame.

Present-day video cameras use arrays of solid state photodiodes (charge coupled devices, or CCDs). Monochrome systems use one CCD chip. Color systems employ three chips—each exposed, via a prism and appropriate filters, to either red, green, or blue wavebands. Each detector within the CCD chip generates an output voltage equivalent to the brightness (radiation energy) of the terrain within the detector's field of view. The number of detectors within the CCD chip determines the pixel density (resolution) of the system. A range of pixel densities is offered on most commercially available systems. A typical solid state video camera with a half-inch CCD chip generates a video frame composed of 400,000 pixels (700 lines per frame). Though it is not possible to determine the number of pixels contained within a 35mm film because the granules of silver halide (black-and-white film) or dye (color film) do not have uniform size, shape, or spatial distribution, a reasonable estimate would be 2.5–3.0 million pixels (Lillesand and Keifer 1994:121). With present technology, video images taken from the same altitude with cameras of the same focal length will have lower spatial resolution than will those generated using 35mm or 70mm photography (Wright 1993). This is likely to change in the near future, as technology for digital High Definition Television (HDTV) doubles the resolution of future video systems (figure 4.10).

To produce accurate terrain color, a video camera makes adjustments to the gain of the blue and red signals (white balance) according to the color "temperature" of the landscape. White balance can be set manually or left to the camera to adjust automatically. Automatic white balance adjustment does, however, mean that like features—such as pine forest—in different video frames (taken during the same flight or at different times) cannot be interpreted as the same vegetation type solely by color. Unlike 35mm photography, which has an aspect ratio (height to width) of 2:3, a video frame has an aspect ratio of 3:4 (Meisner 1986).

Though lower in resolution, aerial videography has the following advantages over 35mm photography:

- Unlike film that must be developed, video images can be viewed in real time or immediately after data acquisition flights.
- Video tapes are much less expensive than buying and developing 35mm film.
- Video tapes have an audio track that permits real-time voice annotation of terrain features during data acquisition.
- Spectral sensitivity of CCD video cameras (from near-UV to near-IR) is broader than either black-and-white or color 35mm photography.
- CCD cameras have good color and geometric fidelity.
- The geographic location (latitude, longitude, and altitude) of the video camera can be recorded on each video frame very easily by attaching a Global Positioning System receiver (see chapter 5) through a character generator.

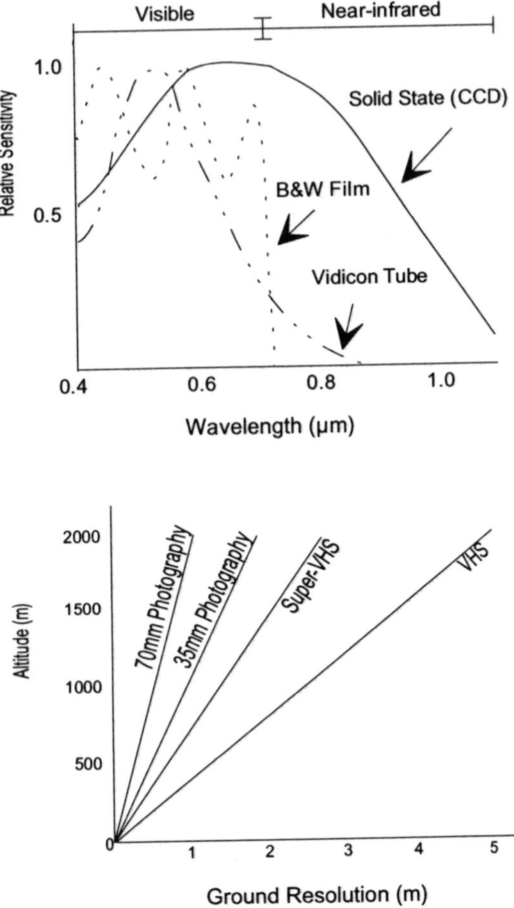

Figure 4.10 Sensitivity and resolution of aerial videography. *Top:* Wavelength sensitivity of vidicon and CCD technology video cameras, relative to black-and-white film. Note how only the new CCD videocameras are sensitive to infrared radiation. *Bottom:* The spatial resolution disadvantages of videography compared to photography intensify as altitude increases. But note that Super VHS videography is a great improvement over the older VHS systems. High Definition Television Systems will further narrow the gap between videography and photography.

- Multispectral image datasets can be generated for digital analysis using a video frame-grabber to capture individual RGB channels from a three-CCD video camera or videorecorder (Everitt et al. 1993).

Aerial Digital Photography

The most recent innovation in digital framing systems merges traditional 35mm photography with CCD digital data collection, in what is called digital photography. The Kodak DCS 200 digital camera features 9×9μm photodiodes in a 1,012×1,524 pixel CCD sensor built into a special back to fit a Nikon 8008s auto focus, auto exposure camera body. The DCS 200 has sufficient RAM to capture a single color or black-and-white image, an optional internal hard drive that can store 50 images, and a SCSI port for downloading images to a laptop or desktop computer. File size is 1.5Mb for a single black-and-white image, and 4.5Mb for a 24 bit color image. Kodak, with Associated Press, developed the NC (News Camera) 200 that uses a 1,280×1,024 CCD, with photodiodes three times bigger than those in the DCS 200—thus providing higher shutter speeds and ASA. The NC 200 uses removable PCMCIA hard drives that are the size of credit cards (105Mb; able to store 75 images) and has a SCSI port to connect to a computer.

Though more expensive than analog photographic imaging, and with somewhat lower resolution, digital photographic cameras offer several distinct advantages. Digital photographic cameras grant almost immediate access to images. The images can be transmitted electronically across telephone lines and computer networks, can be digitally processed and enhanced, and can be integrated easily into documents, using desktop publishing software.

We believe, moreover, that the resolution of digital cameras is likely to match that of 35mm photography in the near future. On the downside, however, as digital camera resolution increases, so too do image storage requirements. For example, the 1,524×1,012 pixel resolution in the DCS 200 requires 1.5Mb to store a single black-and-white image, whereas a 1,732×1,732 resolution camera that matches the resolution of 35mm film would require 3.0Mb to store a single black-and-white image.

5
Getting Started

Formulating the Problem and Determining Imagery Needs
Choosing and Acquiring Images
Deciding Between Analog or Digital Imagery
Acquiring Images
Displaying the Image and Assessing Its Quality
Correcting for Atmospheric Distortions
Correcting for Sensor Errors
Correcting for Geometric Distortions
Image-to-Map Transformation, Using Control Points
Brightness Interpolation
Simple Image-to-Image Coregistration

For remote sensing image analysis to be an effective tool for biodiversity conservation, its use must be problem driven. Without a clearly defined problem, research questions cannot be made explicit, and information needs cannot be characterized. As a result, the most appropriate remote sensing imagery and analysis methods cannot be chosen. To illustrate the critical first steps involved in selecting remote sensing information to solve a resource conservation problem, we will view the process through a hypothetical example: determining the impact of logging on a forested region in Maine. In this way we will be able to demonstrate how to select the most appropriate imagery. Once a source of imagery has been selected, approaches to conducting the analyses are rather similar, regardless of the sensor system being used. Generic operational steps for conducting a study using remote sensing imagery are shown in table 5.1.

Formulating the Problem and Determining Imagery Needs

In our example the question to be addressed is, *What landscape changes occurred as a result of commercial logging in the study area in Maine?*

Table 5.1
Steps in a Remote Sensing Application

1. Formulate the problem
2. Obtain data
    - digitize maps or aerial photographs
    - buy digital imagery
3. Choose image processing system
    - software selection
    - hardware configuration
4. Assess data quality
    - descriptive statistics
    - image display
5. Correct errors
    - radiometric (atmospheric or sensor)
    - geometric
6. Enhance images
    - for digital analysis
    - for visual analysis
7. Conduct field survey
8. Extract features from the imagery
    - classification
    - accuracy assessment
9. Input into a Geographic Information System
    - problem evaluation
10. Summarize results

SOURCE: After Jensen 1986:9.

To answer this question we need to (1) determine the range of landscape types present before logging and (2) be able to measure how the landscape has changed thereafter. Being able to measure how a landscape has changed subsequent to disturbance is, of course, related to the scale of the disturbance and the time that the disturbed areas have had to recover. To conduct a remote sensing analysis on the impact of logging, it is exceedingly important that we have a basic understanding of logging practices (i.e., how logging disturbs a landscape) and of how forest gaps regenerate after different degrees of disturbance (i.e., which if any species invade, how long they are evident in the regenerating gap, whether gaps eventually return to their original species composition, etc.). We cannot overemphasize that a key determinant of the success of studies using remote sensing image analysis is how much one knows about the ecology and land use of the area.

The problem thus formulated, we now need to determine the kind or kinds of imagery to use. From chapter 2 we know that imagery varies according to its spatial, spectral, and temporal resolution, and its area coverage. To identify what form of imagery would best determine the impact of logging on forest habitat we need to answer the following questions:

- What is the minimum feature size to be resolved by the imagery? (This is a function of the size of the logged areas and whether the areas were clearcut or selectively logged.)
- What spectral wavebands would best distinguish logged areas from unlogged areas? (This is a function of the extent of disturbance associated with logging and the degree to which logged areas have recovered.)
- What times of year would yield the best imagery for discriminating the areas affected by logging? (This is a function of the disturbance associated with logging and the degree to which logged areas have recovered.)
- What times of year are likely to exhibit the least atmospheric distortion (haze, clouds, smoke, and dust) and the most appropriate sun angle to minimize topographic shadows?
- How large an area must be surveyed?

Jensen (1986:29) suggests that for a feature to be identified reliably the image of it should comprise at least 20–50 pixels. Our own past experience in monitoring forestry practices (always an important part of any remote sensing study) leads us to decide that the imagery must be suitable for discerning logged areas as small as two hectares. In this instance, logging company records indicate that the landscape was subjected entirely to clearcutting rather than to selective logging. Knowing whether the timber areas were clearcut or selectively logged is critical to deciding on appropriate sources of imagery and analysis methods. Selective logging, although uncommon in the low species diversity boreal woodlands dominated by conifers, is the preferred practice in most tropical forested regions of the world. Selective logging removes only commercially valuable species from an area, leaving most other trees standing. Depending on the density of tree species that have a market, the scale of disturbance associated with selective logging can vary enormously, as can our ability to detect the impact of logging using remote sensing imagery. Selectively logged areas are likely to exhibit more subtle changes in landscape composition and structure than are clearcut areas. Remote sensing imagery, therefore, would have to contain more detailed information (i.e., have higher spatial, spectral, and radiometric resolution) to detect the impact of selective logging on a forest landscape.

We know that clearcutting causes a complete removal of trees from an area, and that regenerating vegetation, at least initially, are different species. The spectral characteristics of unlogged and recently logged areas are likely, therefore, to be quite different. As logged areas recover, our ability to discriminate them from unlogged areas will, of course, diminish with time. Nevertheless, in this case the choice of remote sensing imagery can be based less on the spectral characteristics of the sensor and more on the imagery's spatial characteristics. Logged patches can be as small as two hectares, and to be identified within the imagery, must comprise a minimum of 20–50 pixels. Using the equation,

$$\textit{Spatial resolution} = \sqrt{\frac{\textit{minimum logged area } m^2}{\textit{minimum number of pixels}}}$$

we can calculate that in order to identify the smallest logged area, we need to choose imagery with a spatial resolution no coarser than 20–30m.

Several sources of remote sensing imagery meet our spatial resolution requirements (Landsat TM, SPOT, Soyuzkarta, aerial photography, and videography; see table 5.2). We now need to decide which source of imagery provides the most appropriate spectral resolution for discriminating intact forest from clearcuts in various stages of regeneration, and that has spatial coverage suitable for a state-wide survey.

Both aerial photography and videography are well suited for detailed surveys of relatively small areas. Costs of obtaining and analyzing these two sources of imagery become prohibitive, in terms of time and money, for regional or larger area studies. Satellite imagery offers the best balance between spatial resolution and spatial coverage.

We now must ask which source of satellite remote sensing imagery offers the best spectral information to help answer our research question. To do this we need a basic understanding of how the spectral response of forest is likely to differ from logged patches that may be in stages of vegetational succession anywhere from zero to twenty years post-logging. We know that vegetation and soils have very different spectral reflectance characteristics, as soil reflectance increases with increasing wavelength, whereas vegetation reflects radiation strongly only in the green and IR

Table 5.2

Characteristics of Remote Sensing Satellite Sensor Systems

Satellite	**Sensor**	**Band**	**Waveband**	**Pixel**	**Levels**	**Swath**	**Revisit**
Meteosat	VIS	1	0.4-1.1	2.4km	256		30 minutes
	IR	2	5.7-7.1	2.4km	256		30 minutes
	WV	3	10.5-12.5	2.4km	256		30 minutes
GOES	VISSR	1	0.55-0.72	0.9km	64		30 minutes
		2	10.5-12.6	6.9km	256		30 minutes
NOAA	AVHRR	1	0.58-0.68	1.1/4km	1024	2700km	12 hours
		2	0.72-1.10	1.1/4km	1024	2700km	12 hours
		3	3.55-3.93	1.1/4km	1024	2700km	12 hours
		4	10.5-11.3	1.1/4km	1024	2700km	12 hours
		5	11.5-12.5	1.1/4km	1024	2700km	12 hours
NIMBUS	CZCS	1	0.43-0.45	0.825km	256	1566km	17 days
		2	0.51-0.53	0.825km	256	1566km	17 days
		3	0.54-0.56	0.825km	256	1566km	17 days
		4	0.66-0.68	0.825km	256	1566km	17 days
		5	0.70-0.80	0.825km	256	1566km	17 days
		6	10.5-12.5	0.825km	256	1566km	17 days
MOS-1	VTIR	1	0.5-0.7	0.9km	256	1500km	17 days
		2	6.0-7.0	2.7km	256	1500km	
		3	10.5-11.5	2.7km	256	1500km	
		4	11.5-12.5	2.7km	256	1500km	
	MESSR	1	0.51-0.59	50m	64	100km	
		2	0.61-0.69	50m	64	100km	
		3	0.72-0.80	50m	64	100km	
		4	0.80-1.10	50m	64	100km	
Landsat	MSS	1 (4)	0.5-0.6	79m	128	185 km	16/18 days
		2 (5)	0.6-0.7	79m	128	185 km	16/18 days
		3 (6)	0.7-0.8	79m	128	185 km	16/18 days
		4 (7)	0.8-1.1	79m	64/128	185 km	16/18 days
	TM	1	0.45-0.52	30 m	256	185 km	16 days
		2	0.52-0.60	30 m	256	185 km	16 days
		3	0.63-0.69	30 m	256	185 km	16 days
		4	0.76-0.90	30 m	256	185 km	16 days
		5	1.55-1.75	30 m	256	185 km	16 days
		6	10.4-12.5	120 m	256	185 km	16 days
		7	2.08-2.35	30 m	256	185 km	16 days
IRS-1	LISS1/2	1	0.45-0.52	73/36 m	128	148 km	22 days
		2	0.52-0.59	73/36 m	128	148 km	
		3	0.62-0.68	73/36 m	128	148 km	
		4	0.77-0.86	73/36 m	128	148 km	
JERS-1	OPS	1	0.56-0.64	20 m	64	75 km	44 days
		2	0.66-0.72	20 m	64	75 km	
		3	0.81-0.91	20 m	64	75 km	
		4	0.81-0.91	20 m	64	75 km	
		5	1.65-1.76	20 m	64	75 km	
		6	2.06-2.17	20 m	64	75 km	
		7	2.19-2.31	20 m	64	75 km	
		8	2.33-2.46	20 m	64	75 km	
SPOT	HRV	1	0.50-0.59	20 m	256	60 km	5-26 days
		2	0.61-0.68	20 m	256	60 km	5-26 days
		3	0.79-0.89	20 m	256	60 km	5-26 days
	HRV	PAN	0.51-0.73	10 m	256	60 km	5-26 days
Radar							
ERS-1	SAR		5.7cm C band VV	30m		100km	16-18 days
JERS-1	SAR		23cm L band HH	18 m		75 km	
Radarsat	SAR		5.6cm C band HH	10-100 m		45-500 km	3-24 days

wavebands. To discriminate forest from recently logged areas with exposed soil, we should select a source of imagery that records landscape reflectance (brightness) in at least the green and red wavebands. Panchromatic imagery obtained by the SPOT satellite has suitable spatial resolution (10m) and records landscape brightness in the green to red (0.52–0.73μm) range. A panchromatic, or gray-scale, image does not allow us to determine whether a bright patch of landscape is bright green or bright yellow, orange or red. Multispectral imagery is thus key to helping us discriminate logged from unlogged areas.

We know that unlogged areas in Maine are dominated by conifers, and that the early colonizers of clearcut patches are primarily birches and aspens. Needleleaf and broadleaf trees have very similar leaf reflectance characteristics through the blue, green, and red wavebands, and only start to diverge in the reflected IR wavebands between 0.75 and 3.0μm, where needleleaf trees show less reflectance. To discriminate between areas of unlogged forest characterized by needleleaf conifers and logged areas dominated by broadleaf trees, we should select imagery that records reflected IR. Though the Landsat, SPOT, and Soyuzkarta KATE-200 all have IR detectors, Landsat TM has the largest number of IR waveband channels (three).

Our next decision as to the best source of imagery is based on the availability of imagery at the best time of year for discriminating between logged and unlogged areas. Images obtained in the winter when broadleaf trees are leafless would certainly make unlogged conifer forest spectrally different from patches of logged and regenerating forest dominated by leafless birches. However, logged patches with a lot of exposed soil might look spectrally very similar to fallow farmland, particularly when snow covered. Images obtained in the spring after leaf flush would have few cloud or haze problems, given typical weather conditions. During this season discriminating logged from unlogged areas would have to rely on the somewhat different spectral characteristics of needleleaf and broadleaf tree species. Summer imagery is likely to suffer moderate to severe haze problems, given the typically humid climate of Maine between June and late August. Imagery obtained in the autumn might have few haze and cloud problems, and would benefit from the fact that the green leaves of broadleaf trees turn a spectrally distinctive yellow, orange, and red. The best seasons for obtaining imagery to discriminate logged from unlogged area are therefore autumn and spring.

Soyuzkarta KATE-200 imagery is obtained irregularly during Russian unmanned space missions. No imagery is available for Maine during the spring or autumn months, and would have to be commissioned on a future space mission. A search of the SPOT and Landsat image archives revealed that 1994 or 1995 imagery with less than 10% cloud cover was not available for the whole state of Maine. New images would have to be ordered to complete the state-wide coverage for either remote sensing source. Multispectral imagery from SPOT HRV and Landsat TM can be obtained fifteen to twenty times a year for any location on earth; thus it is possible to order imagery to be taken during the autumn months when broadleaf color is most distinctive. Archival material from 1989 and 1990 is available from both SPOT and Landsat. Because the first SPOT satellite was launched in 1986, older archival material is available only from Landsat. Cloud cover problems in the archival images may require purchasing several images over the same area and cutting and pasting cloud-free areas into a single composite image; this, of course, increases the cost.

The final decision in choosing the optimal source of imagery is based on cost. A single SPOT scene covers only a 60×60km area, compared to the 185×185km coverage of a Landsat scene. Given the price of SPOT and Landsat imagery and the size of each scene, SPOT would be almost five times as expensive as Landsat for a regional survey of logging in Maine. We thus conclude that Landsat TM imagery provides the optimal mix of spatial, spectral, and temporal resolution, spatial coverage, availability, and cost for our particular study.

Using our knowledge of the target area and our understanding of the EMR reflectance characteristics of vegetation, soils, and water as rough guides, we are able to choose a source of imagery that has the spatial, spectral, and temporal resolution appropriate to detecting and identifying the features and phenomena that are relevant to our problem, at a cost we can afford.

Choosing and Acquiring Images

Though aerial photography still provides the highest spatial resolution remote sensing imagery that is readily available, digital imagery is quickly becoming the most important source of imagery for regional-level studies of natural resources and rapid environmental assessments. As several excellent texts are in print that describe the intricacies of aerial photo interpretation (Lillesand and Kiefer 1994; American Society of Photogrammetry 1980), we will focus on describing how to acquire and use digital remote sensing imagery.

Deciding Between Analog or Digital Imagery

Once we have decided that satellite imagery is likely to provide the best mix of area coverage and spectral, spatial, and temporal resolution, we need to determine whether to obtain photographic (analog) or digital imagery. Photographic data for most satellite systems are available as paper prints, negatives or positive transparencies of single spectral bands in black-and-white, or 3-band composites in color. Digital data from all spectral bands of an image are distributed on magnetic media such as diskettes (720k–2.88Mb), 1,200ft tapes (23Mb at 1,600bpi, 90Mb at 6,250bpi), compact 8mm helical tapes (2–5Gb; 1Gb=1,000Mb), or more recently on optical discs (CD-ROM; 650Mb).

Though analysis of remote sensing imagery traditionally has been accomplished solely by the visual interpretation of aerial photographic data, information from satellite imagery is extracted most effectively using a combination of quantitative digital analysis and visual interpretation methods. Visual interpretation is still the most effective way to extract spatial information such as the shape, orientation, juxtaposition, and texture of features. Visual interpretation is, however, largely qualitative and cannot make use of the full range of multispectral and radiometric information contained within imagery. Thus, though all remote sensing analyses rely on the visual interpretation skills of a human analyst, digital processing of the imagery can substantially enhance important features for subsequent visual interpretation. Analysis of multispectral imagery can be achieved using either digital or visual interpretation. However, it is most effective using a combination of both approaches. Table 5.3 compares the attributes of both approaches.

Photographic remote sensing images that were generated from digital data can be converted back into digital form using a scanner. However, information is usually lost in the process. Thus, although digital imagery is usually more expensive than photographic (analog) imagery, it contains more information and is considerably more flexible. Obtaining remote sensing imagery in digital form greatly improves the success of visual interpretation by allowing for computer enhancement of important features

Table 5.3
Advantages and Disadvantages of Visual Versus Digital Analysis of Imagery

Visual Interpretation	Digital/Computer Processing
Spatial information, such as shape, size, texture, pattern, location, orientation, and association, is easily determined qualitatively.	Few methods are available to analyze spatial patterns, orientations, and associations; shape determination requires complex software; size (area) is readily achieved quantitatively.
Human eye is only sensitive to about 30 levels of brightness.	Can quantitatively analyze all radiometric detail/brightness levels (e.g., 64, 128, 256, 512, 1024; 6–10 bits respectively).
Limited multispectral (color) discrimination, as the color of a feature is affected by the background color. Each color image is usually composed from only three spectral bands.	Can analyze multidimensional spectral information quantitatively, and can analyze any number of bands simultaneously.
Multiple pixel level of analysis.	Individual pixel level of analysis. Neighborhood analysis is available on some systems.

SOURCE: After Richards 1986:70.

and reduction of distracting noise. *We suggest that whenever budgets and equipment permit, users should always opt for obtaining remote sensing imagery in digital form, even when information within the imagery is ultimately to be extracted by visual interpretation.*

Once the imagery has been selected, the next step is to decide what format to purchase—prints, transparencies, or digital data. Your choice of format will be determined by the size of the area of interest and your hardware and software configuration.

Acquiring Images

Once the type of imagery has been selected, we need to determine whether such imagery is indeed available for the area in question during a suitable time of year for distinguishing features of interest from the background.

Satellite imagery is supplied primarily by the distributors listed in table 5.4. These distributors also provide lists of companies that sell value-added products and services using satellite imagery—for example, geometrically (map) corrected and digitally enhanced imagery. To order imagery from these distributors, we must request a search of their archives to determine if suitable imagery exists. Distributors do not charge a fee for the search. To conduct a search, the following information must be provided:

- latitude and longitude coordinates of the area's corner points;
- preferred dates (e.g., certain seasons or years, no older than *x* years, etc.);

Table 5.4
Distributors of Commonly Used Digital Forms of Remote Sensing Imagery

Landsat TM	EOSAT, 4300 Forbes Boulevard, Lanham, Maryland 20706, USA Tel 1-301-552-0571 FAX 1-301-552-3762
Landsat MSS	EROS DATA CENTER, USGS, Sioux Falls, South Dakota 57198, USA Tel 1-605-594-6151
SPOT Panchromatic and HRV	SPOT IMAGE, Cedex 16 bis, av. Edouard Belin, 31030 Toulouse, France Tel 33-61-53-99-76 FAX 33-61-28-18-59
Soyuzkarta photography	SPOT Image Corp., 1897 Preston White Drive, Reston, Virginia 22091, USA Tel 1-703-620-2200 FAX 1-703-648-1813
JERS-1, MOS 1,2	Remote Sensing Technology Center of Japan (RESTEC), Uni-Roppongi Bld., 7-15-17 Roppongi, Minato-Ku, Tokyo 106, Japan Tel 81-03-402-1761 FAX 81-03-403-1766
IRS-1	National Remote Sensing Agency (NRSA), Department of Space, Balanagar, Hyderabad, 500 037, Andhra Pradesh, India Tel 0425-6522
ERS-1	Eurimage, Viae E. D'Onofria 212, 00511 Rome, Italy Tel 39-6-406941 FAX 39-6-406942
Almaz SAR	Hughes STX Corp., 4400 Forbes Blvd., Lanham, Maryland 20706, USA Tel 1-301-794-5020 FAX 1-301-306-0963

- format of data required (multispectral, single band, full scene, quarter scene, subscene window); and
- maximum percentage of cloud cover acceptable (usually less than 30%)

The distributor will return to the applicant a list of all available imagery, specifying the date acquired, area coverage, and percentage cloud cover. We can then select the most suitable imagery and request a preview or quick-look image. Quick-look images are usually provided free of charge. They are small overviews that show, primarily, the location of clouds. If an archive search fails to retrieve any imagery over our area of interest, we can ask for an acquisition to be made that satisfies our stated location and cloud cover criteria. This may take several weeks to months, depending on the revisit frequency of the satellite and weather conditions at the study location.

Displaying the Image and Assessing Its Quality

One of the first things that you will want to do after receiving the digital imagery is to display it. Imagery can be displayed on a computer, either one band at a time in black and white or in three-band combinations called *color composites*. Young's additive theory of color states that when the three primary colors of light (red, green, and blue) are combined, they can produce all the hues in the visible spectrum (plate 5). Thus, by assigning each spectral band in a multispectral image to be represented by the colors blue, green, or red, the resulting composite image can exhibit all the colors of the visible spectrum. As only three primary additive colors exist, only three bands of a multiband image can be displayed simultaneously.

Selecting which three bands of a multispectral image to combine in a composite image is mostly a matter of experience and trial and error, because each application attempts to emphasize different features within the imagery. A quantitative method to maximize the "information" content of a three-band composite was developed by Chavez et al. (1982). This optimum index factor (OIF) allows all three band combinations to be ranked according to the total variance and correlation within and between band combinations. The OIF for each three-band combination is calculated for three spectral bands, i, j, and k, using this equation,

$$OIF_{(i,j,k)} = \frac{s_i + s_j + s_k}{|r_{ij}|+|r_{ik}|+|r_{jk}|},$$

where s_k is the standard deviation for band k, and $|r_{ij}|$ is the absolute value of the correlation coefficient between bands i and j. The three-band composite with the largest OIF will have the most information (highest variance) with the least redundancy (lowest correlation). Once the highest ranked composite is identified, you must still decide which color (blue, green, or red) to assign to each band to emphasize the features of interest. Table 5.5 shows the OIF calculations for a portion of a Landsat TM image of an area in Maine. Plate 6 shows color composite images of the three best band combinations, according to the OIF analysis (three-band combinations 134, 145, 345), as well as two images for the worst combination (three-band combinations 236, 126).

Principal components analysis (discussed later in this chapter) can be

Table 5.5
Calculation of OIF for Landsat TM Subscene

Standard Deviation									
Bands	1.000	2	3	4	5	6	7		
	7.887	5.11	7.80	25.80	21.07	4.75	9.04		
Correlation Matrix									
Bands	1.000	2	3	4	5	6	7		
1	1.000								
2	0.901	1.00							
3	0.899	0.94	1.00						
4	0.047	0.19	0.00	1.00					
5	0.473	0.69	0.56	0.76	1.00				
6	0.722	0.80	0.76	0.21	0.66	1.00			
7	0.777	0.91	0.84	0.37	0.85	0.82	1.00		
OIF(123)=	7.603	OIF(234)=	34.197	OIF(345)=	41.546	OIF(456)=	31.665	OIF(567)=	14.956
OIF(124)=	34.057	OIF(235)=	15.575	OIF(346)=	39.178	OIF(457)=	28.256		
OIF(125)=	16.497	OIF(236)=	7.083	OIF(347)=	34.837	OIF(467)=	28.211		
OIF(126)=	7.341	OIF(237)=	8.160	OIF(356)=	16.961				
OIF(127)=	8.512	OIF(245)=	31.750	OIF(357)=	16.867				
OIF(134)=	43.617	OIF(246)=	29.790	OIF(367)=	8.906				
OIF(135)=	19.077	OIF(247)=	27.058						
OIF(136)=	8.577	OIF(256)=	14.382						
OIF(137)=	9.814	OIF(257)=	14.374						
OIF(145)=	42.942	OIF(267)=	7.490						
OIF(146)=	39.258								
OIF(147)=	35.641								
OIF(156)=	18.136								
OIF(157)=	18.106								
OIF(167)=	9.359								

used to reduce a multidimensional image into three components that retain over 95% of the information within the original dataset. However, as this is a synthetic image, we cannot use our knowledge of the reflectance characteristics of landscape features to help interpret the image. When the visible bands of multispectral images are displayed in three-band combinations (TM bands 3, 2, 1 as RGB), the resulting images are referred to as pseudo true-color composites. This is because the broad range of the spectrum detected by each band within most multispectral systems can only provide a very rough approximation of the true ranges of colors evident within a landscape. When nonvisible bands are combined with visible bands (MSS bands 4, 5, 7; TM bands 2, 3, 4), the images produced are known as false-color composites. (Plate 6 shows examples of false-color composites.) The most common false-color composites created from satellite imagery are those that attempt to mimic the look of traditional aerial IR photography, and are generated by displaying the green band as blue, red as green, and VNIR as red (table 5.6).

Not only is it impossible to display all bands of a multispectral image simultaneously, most 1995 computer systems are only capable of displaying images that are smaller than 1,024×768 pixels (the resolution of the screen). As a single Landsat MSS scene contains approximately 2,340×3,240 pixels, every pixel in the image cannot be displayed at once. Two options exist to remedy this. The first is image reduction, where we view only a certain percentage of pixels (i.e., we display every 4th, 8th,

Table 5.6
Color of Landscape Features in a Landsat MSS Band 4, 5, and 7 False Color IR Composite Image

Land-cover	False color
Vegetation	red
Clear water	black or dark blue
Silty water	light blue
Bare soil	blue
Cities	blue
Dry sand	white or yellow
Red rocks	yellow
Clouds/snow	white
Shadows	black

SOURCE: Sabins 1986:81
NOTE: Also TM bands 2, 3, 4.

or 10th pixel). In doing this, we can retain the large area view of the whole image, but we lose detail (resolution) as a result. Image reductions are not usually used for interpretation but more for regional orientation and selection of smaller study areas. The alternative to image reduction is to cut the image into a set of smaller windows, each of which can be displayed at full resolution but, of course, without the synoptic view of the whole image (figure 5.1).

In many cases, particularly in visual interpretation of the data, it is useful to magnify (or zoom in on) portions of the image. Magnification is achieved simply by duplicating individual pixels into a square block of pixels with the same brightness value. Figure 5.2 shows an area of a single-band Landsat TM image of a portion of Maine magnified 1, 2, 4, and 16 times. As the image is magnified, details initially become more visible. Eventually, however, the magnified image takes on a blocky appearance, where individual pixels become visible. At this level of magnification our ability to identify features visually actually declines. This effect of over-magnification is used in the popular illusion shown in figure 5.3, where the white and black blocks are only identifiable as a portrait of President Lincoln when the image is held at arm's length.

A major problem in image processing is producing paper or transparency copies that match exactly the colors within an image as viewed on screen. Color printers use several different technologies (inkjet, thermal wax, sublimation dye), but all use a subtractive color scheme, where cyan, magenta, and yellow inks or dyes are combined to create all colors in the visible spectrum. As computer display systems use an additive color scheme (red, green, and blue), the colors generated by the two systems are exceedingly difficult to match exactly, although vendors such as Tektronix, Apple, and Pantone do sell color matching software that attempts the closest possible match between display and hardcopy devices. To give an example of the difference between the two coloring schemes, if one combines equal amounts of each primary color, the subtractive technology produces black, and the additive white. Color laser imagers (e.g., Cirrus Technology, Nashua, New Hampshire) avoid the mismatch by using an additive color system. Colored lights rather than inks or dyes expose positive transparencies that match exactly the colors seen on screen.

Viewing the imagery on screen is the first and often most effective means of identifying problems, such as cloud cover, haze, and sensor errors (plate 7). However, as you would when beginning the examination

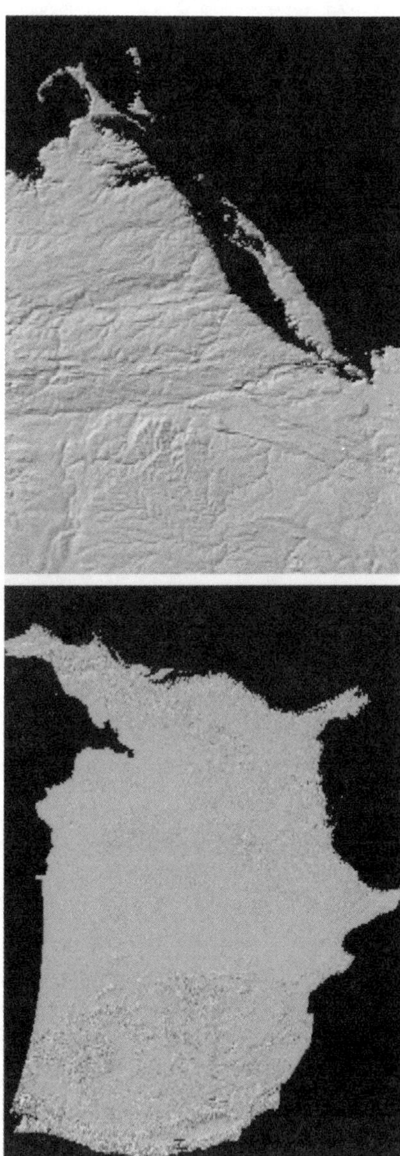

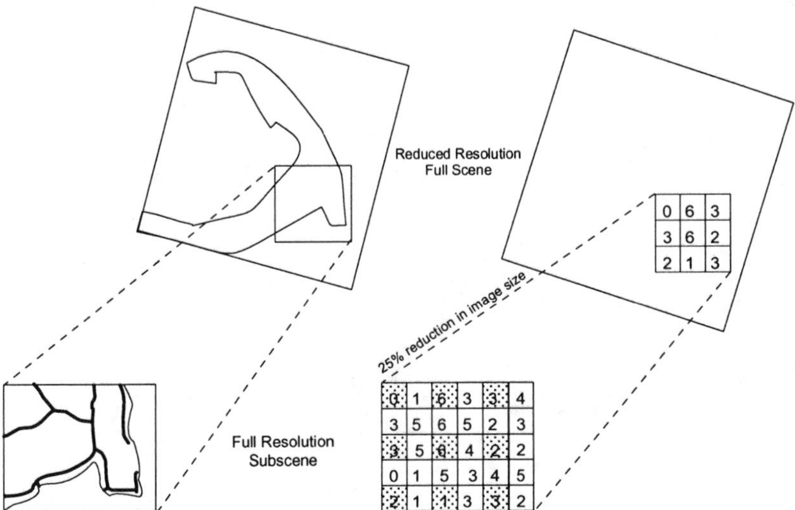

Figure 5.1 Two options for coping with screen display limitations: full scene with reduced resolution or full resolution of a subscene. When the number of pixels in an image exceeds the number of pixels in a computer display, the image can be viewed in its entirety only by displaying every *n*th pixel. Using image reduction, regional- or continental-scale images can be viewed at one time (*left image*), but the elimination of pixels results in a loss of detail. An alternative is to display all pixels within a small section (*right image*) of the full scene. Subscenes lose the synoptic view of a full scene but retain all the spatial resolution (detail). The schematic shows that if one in every four pixels is eliminated (shaded), the image can be reduced to 25% of its original size, and thus four times the area can be displayed (at lower spatial resolution) on the computer screen at one time.

of any numeric research data, it is worthwhile plotting frequency histograms of raw brightness values of each sensor band within the imagery, and conducting exploratory univariate and multivariate statistical analysis. Results of this preliminary analysis will not only help to assess the quality of the data, they are invaluable for displaying, correcting, and enhancing the imagery effectively.

Figure 5.4 shows a window of raw data for Landsat TM band 1 and a histogram of pixel brightness values. The abscissa for the histogram is the range of brightness values (0–255, 8 bits for Landsat TM data). The ordinate is the number of pixels displaying each discrete value. By exam-

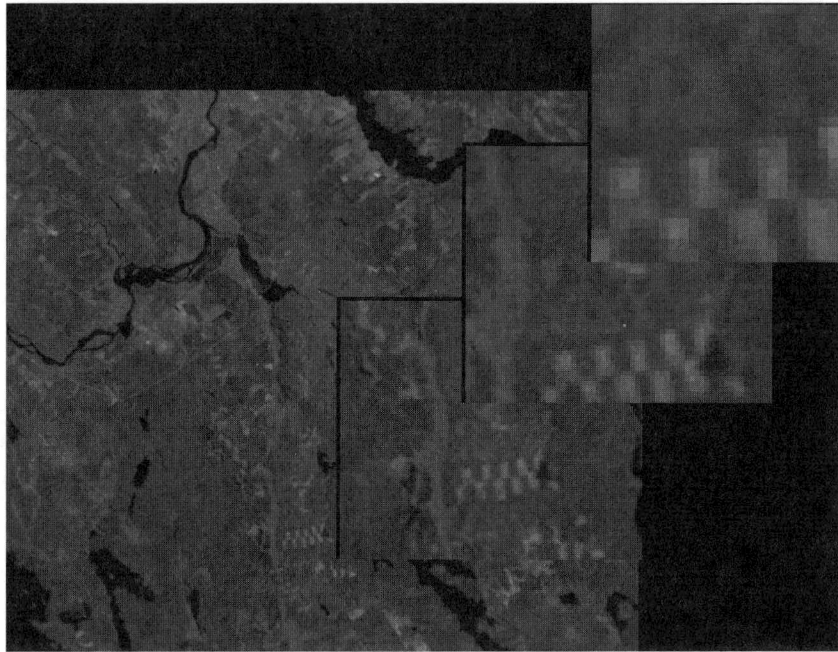

Figure 5.2 Single-band Landsat TM image magnified 1, 2, 4, and 16 times. Notice how the very blocky appearance of the 16× magnified image actually diminishes our ability to see details relative to, say, the 2× magnified image.

ining raw data histograms, such as this example, we can very quickly see the range (contrast) and distribution (unimodal or multimodal) of brightness values. (Figure 5.4 displays unimodal brightness distribution and low contrast.)

A very narrow range of brightness values (low contrast) within an image may indicate homogeneity of the landscape. More often, low contrast is a warning that the image quality is impaired by atmospheric interference (haze) or limited sensor sensitivity. A narrow range of brightness values (as in figure 5.4) may predict difficulties in distinguishing relevant landscape features. Displaying raw data histograms also allows us to determine, visually, the effect of any subsequent data transformations performed on the imagery.

Calculation of the mean, standard deviation, and maximum and minimum of the brightness values provides a quantitative analysis of the information contained within the imagery that was first assessed more qualita-

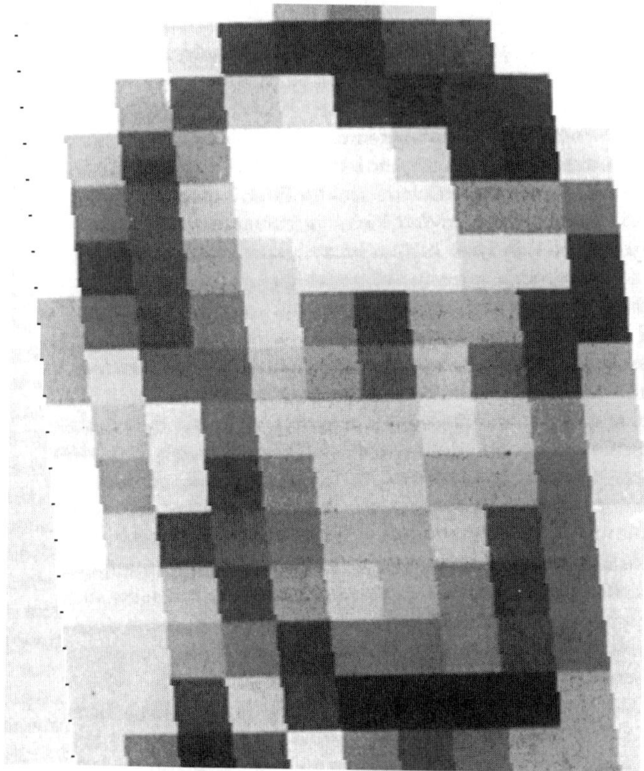

Figure 5.3 Over-magnification hinders interpretation. This pixel portrait is easily identified as Lincoln only when we reduce its magnification by holding it at arm's length.

tively in a brightness histogram or when displayed on screen. Table 5.7 presents these calculations for spectral bands 1 to 7 within a small window of a Landsat TM image located over Calais, Maine. The Landsat TM window is 67 columns wide by 281 rows deep, and is composed of 18,827 individual pixels. Infrared bands 4 and 5 have the highest contrast as measured by the standard deviation of pixel brightness values. Thermal IR band 6 and visible green band 2 have the lowest contrast.

More important, these calculations are useful for determining whether the brightness value of one band changes predictably with that of another band. If an increase in the brightness of pixels in band 1, for example, is matched by an increase or decrease in the brightness of corresponding pixels in another band, then the data are not independent; the informa-

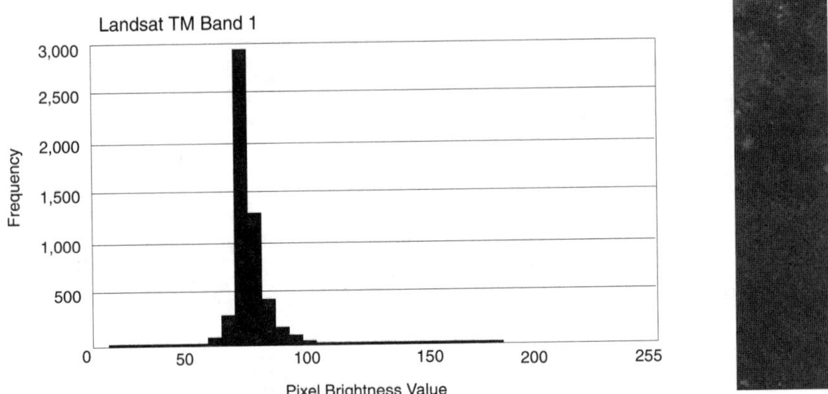

Figure 5.4 A low-contrast, unenhanced image of Landsat TM band 1 and its brightness histogram. Even though pixels range in brightness from 3 to 184, most pixels fall within a few brightness values. As a result, the tone of the image is dominated by a few brightness values and thus shows little contrast.

Table 5.7
Descriptive Statistics for Raw and Stretched Spectral Bands

Image - Window	Minimum	Maximum	Mean	Standard Deviation	N
Band1	3	184	72.2237	7.89	18827
Band2	19	95	28.8936	5.11	18827
Band3	11	117	24.3952	7.80	18827
Band4	13	225	88.8029	25.80	18827
Band5	4	171	63.5183	21.07	18827
Band6	120	149	130.2094	4.75	18827
Band7	1	144	19.5893	9.04	18827
Stretched spectral bands					
Band1 min-max linear	0	255	24.8121	15.81	18827
Band1 histoequalization	0	255	139.5511	68.72	18827
Band1 5% saturation	0	255	59.5477	70.24	18827
Band4 min-max linear	0	255	91.0457	31.16	18827
Band4 histoequalization	0	255	129.6405	74.24	18827
Band4 5% saturation	0	255	183.9261	65.49	18827

NOTE: Calculations based on data in table 5.4.

tion contained within the two bands is to some extent redundant. To determine the amount of information overlap, we calculate the covariance and correlation between pairs of bands within the imagery. Correlation is a unitless measure of interrelation of two spectral bands. Correlation ranges from $+1$ to -1, where 0 indicates no linear relationship between the two bands.

Examining the correlation matrix for the Maine Landsat TM scene (table 5.5), we see that bands 2 (green) and 7 (IR) are highly interrelated. We know that the area covered by the image is primarily vegetated. Figure 5.5 shows that the spectral ranges of bands 2 and 7 are both in the spectral regions in which vegetation is characteristically reflective. We should therefore expect these two bands to be correlated. To reduce costs and analysis time, we might consider selecting only one band for subsequent analysis, as much of the information in one band is duplicated in the other.

Successful analysis of any research dataset is related to the depth of the analyst's understanding of the information contained within the dataset. It is thus exceedingly important that users of remote sensing information do not skimp on exploratory viewing and statistical analyses of their imagery. The more one understands the spectral characteristics of the imagery, the more able one is to extract useful information from the data.

Correcting for Atmospheric Distortions

Before the imagery is used to address ecologically relevant questions, errors introduced during the imaging process must be removed. The corrected image will then geometrically and radiometrically represent the landscape as realistically as possible.

Image restoration attempts to counter errors generated by the sensors themselves, by atmospheric scattering, and by geometric distortions associated with satellite movement, earth's curvature and rotation, and sensor scan rate. In the remaining sections of this chapter, we will focus on correcting major sources of error.

The primary causes of radiometric (brightness level) errors in multispectral satellite imagery are attributable to (1) atmospheric problems or (2) the sensor instruments themselves. In this first section we examine the atmospheric causes of error; in the next, the equipment causes.

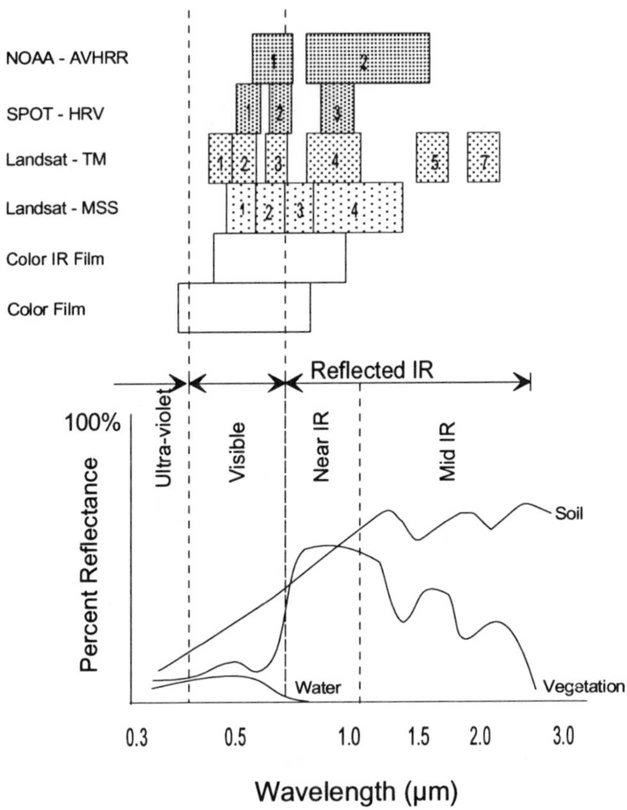

Figure 5.5 Spectral response of landscape features and correlation of spectral bands. The waveband location of Landsat TM bands 5 and 7 were designed to correspond to two mid-IR reflectance peaks of vegetation. As a result, in vegetation-dominated landscapes (such as the image over Calais, Maine) the brightness values of pixels in bands 5 and 7 are often highly correlated. Thus, depending on the research question, one could eliminate either band 5 or 7 to economize on image storage costs and computation time without jeopardizing interpretive accuracy.

Passive remote sensing systems, such as Landsat and SPOT, record the visible and infrared radiation that reflects or is emitted from land and sea surfaces. Without an atmosphere the brightness values recorded by satellite sensors would be a simple function of the solar energy striking the landscape within a given pixel, and the reflective characteristics of that landscape. Atmospheric molecules—notably, oxygen, ozone, carbon dioxide, and water—absorb radiation very strongly along certain wavelengths. Sensor systems are, however, designed to avoid these regions. The remaining atmospheric effects that impinge on the data gathered by sensors are largely the result of light scattering off

- air molecules (Rayleigh scattering);
- particles associated with smoke and dust (0.1–10μm radius; Mie scattering); and
- water droplets (>10μm radius; nonselective scattering).

On a clear day Rayleigh scattering of short wavelength visible light causes the sky to appear blue, regardless of where we look. Similarly, at dusk the long atmospheric path between the setting sun and an observer results in all blue light being scattered away by the time it reaches the viewer, causing sunsets to appear orange and red. Mie scattering also affects shorter wavelengths, but not to the same extent. In contrast, clouds, haze, and fog appear bluish-white or white because large particle scattering is not wavelength specific. Nonselective scattering causes objects to lose color and contrast (the ratio of brightest to darkest areas of a landscape) with increasing distance from the observer or sensor (figure 5.6). Artists take advantage of this effect to convey depth on a two-dimensional canvas. The downside of nonselective scattering is that it reduces detail and thus our ability to detect useful features within imagery.

If you happen to know the temperature, relative humidity, atmospheric pressure, and visibility in your area of interest at the exact time that the satellite image was recorded, it is possible to develop a function for each sensor waveband to correct the absolute and differential atmospheric errors incorporated in the brightness values recorded for each pixel within the image. For most applications this precision of error correction is unnecessary; more often than not, the atmospheric data are not available. As a result, a simplified, more generalized, and admittedly approximate approach must be used to remove atmospheric scattering effects.

Figure 5.7 shows an image for each of the seven spectral bands recorded by Landsat TM near Calais, Maine; figure 5.8 presents the brightness histogram for each. It is reasonable to expect that in each band of the same image some pixels should have brightness values at or close to zero (e.g., areas of clear water or deep shadow). But notice that the minimum values in the histograms of figure 5.8 and in table 5.7 are usually

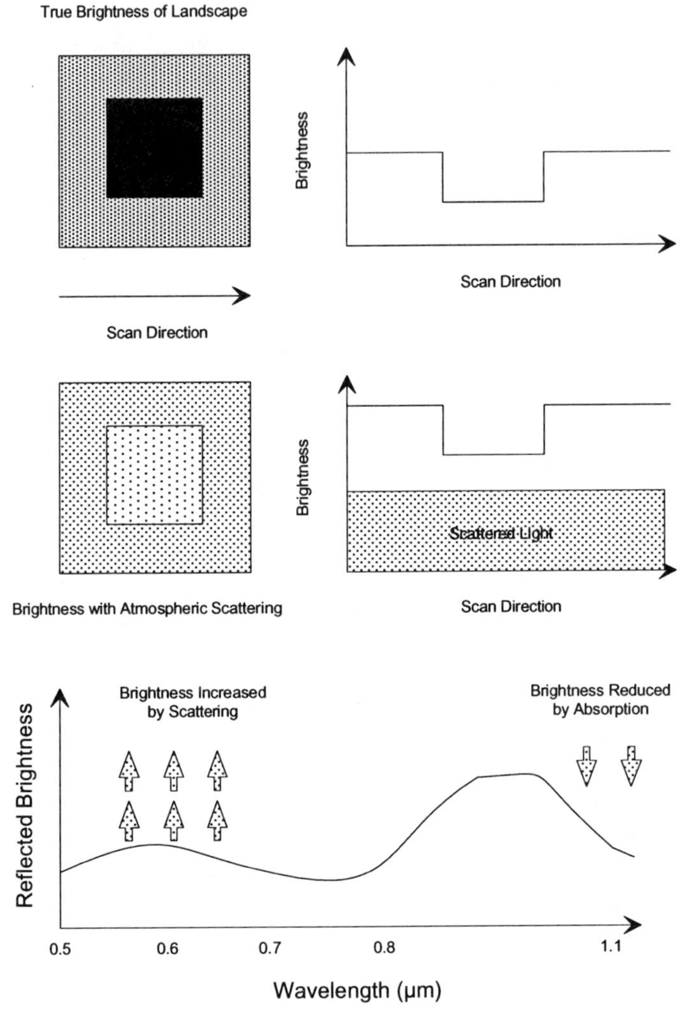

Figure 5.6 Effect of the atmosphere on the contrast ratio of an image.

greater than zero. This is evidence that atmospheric particle scattering has added brightness to each pixel within a band.

Because Rayleigh and Mie scattering affect short (blue) wavelengths the most, we should expect that the visible blue and green bands are "shifted" further from zero than are the red and IR bands. To correct these errors in brightness, we must first determine for each band the size

Figure 5.7 Raw (unenhanced) image at Landsat TM bands 1 to 7. The images are arranged sequentially, with band 1 on the left.

of the histogram "shift," then subtract that amount from the brightness level of each pixel. This form of atmospheric correction is often referred to as *haze removal.* When the bluish-white haze evident in a visible-band color composite image is removed, the color contrast of the image improves markedly.

A more complex adjustment regresses brightness values of pixels selected from an area of homogeneous shadow or deep water (both of which should have brightness values of near zero) in a visible band against the same area in an infrared band. If no atmospheric scattering occurred, the intercept of the regression line should pass through the origin (zero on both x and y axes). If it does not, then each pixel in the visible band should be adjusted by the amount that the intercept is shifted from the origin. This procedure is then repeated for all other bands within the image. We should note that this method is more time consuming, yet it is often no more effective than the simple subtraction adjustment described previously.

These generalized haze removal methods all operate on the assumption that the haze problem is uniform over the entire image. We thus feel confident in lowering the brightness value of each and every pixel within that image. If, however, the haze only occurs in patches over the image, haze

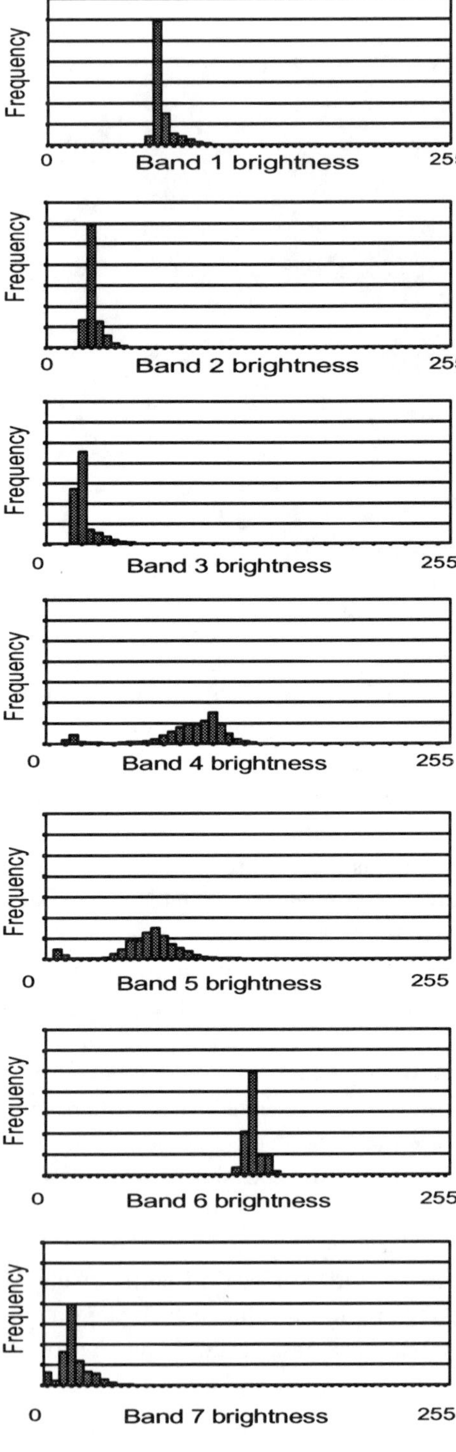
Frequency
0
Band 1 brightness
255
Frequency
0
Band 2 brightness
255
Frequency
0
Band 3 brightness
255
Frequency
0
Band 4 brightness
255
Frequency
0
Band 5 brightness
255
Frequency
0
Band 6 brightness
255
Frequency
0
Band 7 brightness
255

removal that acts on all pixels will erroneously alter the brightness of nonhazy areas. To overcome the problem of uneven distribution of haze within an image, it is possible to mask areas with little or no visible haze. The atmospheric correction can then be completed on only those areas obscured by haze.

In most remote sensing software packages it is fairly easy to create an *image mask*. The process usually involves

1. displaying the image on screen;
2. drawing boundaries around the areas (polygons) that are hazy;
3. saving the polygons in vector format;
4. creating a new image of the same dimensions as the original;
5. inserting the vector polygons as raster areas within the new image, where pixels within the haze polygons are assigned a value of 1 and all other pixels a value of zero; and
6. multiplying each pixel within the binary haze mask image by its corresponding pixel in the original image.

The resulting masked image will display only the hazy regions, as all other areas will be blanked out (i.e., set to zero).

To restore the image, a *reverse mask image* is created. In this the haze areas are blank (zero), and all other areas are displayed with their original brightness values. The masked image and the reverse mask image are then added together to recombine the hazeless and haze-removed areas of the image.

Correcting for Sensor Errors

In scanning mirror systems, such as Landsat MSS and TM, and in the pushbroom system used by SPOT, multiple detectors gather brightness information simultaneously for each spectral band. If the 6 detectors in Landsat MSS, 16 detectors in Landsat TM, or 6,000 in SPOT panchromatic (3,000 in multispectral) are not calibrated relative to one another, then the contrast of the band will be irregular and striping or banding will be evident in the image (figure 5.9). If a detector fails, the error, often

---

Figure 5.8 Brightness histogram for each band in the Landsat TM image shown in figure 5.7.

Figure 5.9 Example of horizontal striping. Unequal calibration of the six sensors in Landsat MSS band 2 result in the horizontal striping seen in this image of the Patuca River in eastern Honduras.

named *line drop,* will appear as single or regularly spaced black lines in the image.

To correct line drop, we need to determine which detector is bad. We begin by identifying each scan line that has a mean brightness at or near zero. Once these lines have been located, we can interpolate a value for each missing pixel along each bad scan line by averaging the brightness values for adjacent pixels in the preceding and succeeding lines.

Correcting miscalibrated sensors is somewhat more complex. Perfectly calibrated detectors would exhibit the same linear relationship between the radiation level falling on the detector and the brightness value recorded (referred to as transfer characteristics; figure 5.10). In reality this is, of course, not the case, and all detectors within the same satellite (say, the sixteen detectors in Landsat TM) show different transfer characteristics among themselves.

To correct poorly adjusted detectors that produce visible banding or striping in an image, one must assume that the distribution of brightness levels for all detectors within a band should be very similar when exam-

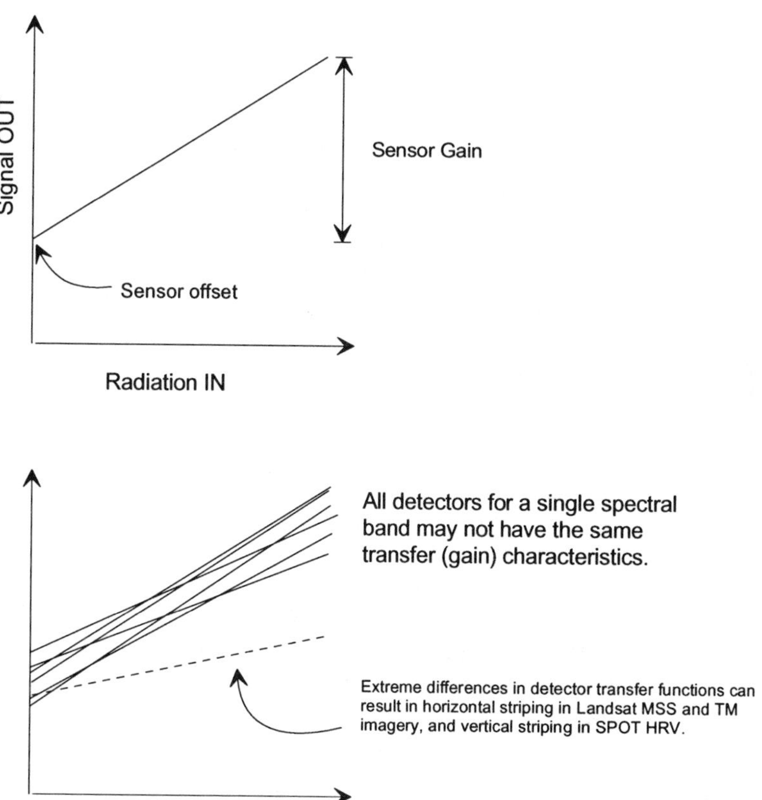

Figure 5.10 Hypothetical radiometric gain and offset characteristics of sensors within a radiation detector. After Richards 1986:37.

ined over the whole image. Given the generally random distribution and orientation of features over a landscape, this assumption is usually reasonable. We can thus correct scan lines with noticeably brighter or darker pixels by adjusting the brightness values to match the mean (or median) and standard deviation of one of the other detectors (acting as the standard), or of all other pixels combined. It is of course possible to statistically compare all detectors in all bands and correct those that show significant differences. However, if striping is not visible, this procedure is usually not necessary.

In some cases, such as with the Landsat TM image over Calais, Maine, (plate 7), random pixels throughout individual bands are either *dropped*

Table 5.8
Moving Window Algorithm for Removing Random Pixel Errors

BV1	BV2	BV3
BV4	BV5	BV6
BV7	BV8	BV9

Algorithm for pixel drop removal:

Mean(a) = (BV1 + BV3 + BV7 + BV9) ÷ 4
Mean(b) = (BV2 + BV4 + BV6 + BV8) ÷ 4
Diff = Mean(a) − Mean(b)
Threshold = Diff × Weight

IF (BV5 − Mean[a]) OR (BV5 − Mean[b]) > Threshold
  BV5 = Median(b)
ELSE
  BV5 = BV5
NB: The higher the weight, the fewer the pixels that will exceed the threshold.

SOURCE: Adapted from Lillesand and Kiefer 1994:540.

(have a value of zero) or *saturated* (have a value of 255). This creates a speckled or salt-and-pepper look to the image. To remove random pixel errors, the most effective technique is to apply a median filter (discussed later) only to pixels with a brightness value above or below a threshold (table 5.8). A simple median filter (using a 3×3 window), when applied to all pixels in an image, will remove random noise pixels, but it will also bring about a smoothing of the image and some loss in visual sharpness.

Correcting for Geometric Distortions

Although radiometric errors are much more easily seen within raw satellite imagery, geometric distortion is often more severe, and correction of errors is certainly more complex. Failure to remove geometric distortion: (1) limits our ability to relate spectral features distinguishable in the image with landscape features on the ground; (2) can make area estimates almost worthless; (3) precludes integration of imagery with other sources of map data that conform to a standard map projection (the representation of a curved surface on a flat sheet); and (4) prevents two images from different dates from being compared pixel by pixel. Major sources of geometric distortions include

- earth's rotation under the satellite during image acquisition (figure 5.11),

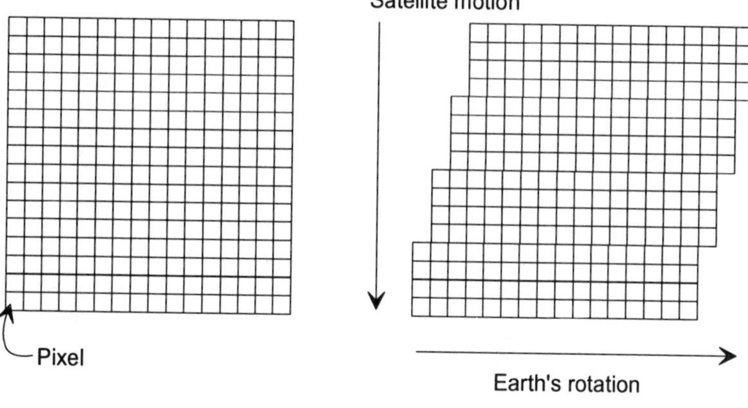

Figure 5.11 Scan line offset of an image, attributable to earth's rotation under the satellite. *Earth's rotation* is a major source of geometric distortion. Near-polar orbiting satellites, such as Landsat and SPOT, gather brightness data as they travel southward over the landscape. At the same time, the earth is rotating to the east. Thus, the end of a line of data collected at the southern edge of an image is not parallel to the line on the northern edge. Each successive line (SPOT) or block of lines (6 for Landsat MSS, 16 for TM) is shifted to the west by a distance related to the frame (image) scan speed of the satellite and the latitude of frame acquisition. At Boston (latitude 42.36°), the earth moves 9.8km during the 28.6 seconds required to scan the 185km frame. This represents about 5% of the image frame. Thus each block of TM lines must be offset to remove the skew (5%) introduced into the imagery. Surface velocity is calculated as the product of the cosine of latitude, earth's radius [6.37816 Mm], and earth's rotational velocity [72.72 μrad s^{-1}]. Source: Richards 1986:44.

- earth's curvature,
- panoramic effects (view angle),
- topography,
- gravitational anomalies causing speed and altitude changes in the satellite, and
- satellite platform instability.

There are two approaches for correcting geometric distortions. The first attempts to model the nature and magnitude of all sources of distortion and generate correction formulae (Mather 1987). This approach works well when the sources of distortion are well understood and uniform in their effect. As we know little about the unsystematic variations in satellite altitude and attitude, however, this method can never remove all distortion. The second approach uses a set of ground control points within an image (whose true geographic coordinates are known) to statistically generate two functions that transform pixel locations within the raw (uncorrected) image into their corresponding coordinates within a standard projection system, such as Universal Transverse Mercator (UTM). This mapping of the image to a standard coordinate system can correct geometric distortions without the user knowing either the source or magnitude of the errors. This second approach is thus the most frequently used procedure, and can achieve within-pixel levels of precision.

Image-to-Map Transformation, Using Control Points

To transform pixel locations—row (r), column (c)—to true map coordinates (x, y) and vice versa, we need to develop two functions:

$$r = a(x,y)$$

$$c = b(x,y)$$

where a and b are functions that account for the geometric distortion within the image.

This geometric correction is achieved by first identifying a set of *ground control points* (GCPs) that can be located unambiguously both in the raw imagery and on a reference map (figure 5.12, table 5.9). Not all reference maps are suitable for establishing the exact location of GCPs. In very small-scale maps, say 1:1,000,000, the lines used to represent roads and rivers may be 200m or more wide. The estimated geographic position of a GCP located at a crossroad might vary by several hundred meters depending on which side of the map line the position is determined. An error of several hundred meters may not be significant when using 1km resolution AVHRR data, but it is problematic when using 10m SPOT imagery. If suitable scale maps are not available, GCPs identified in the image can be visited in the field and their true geographic location determined using a Global Positioning System (GPS) receiver (to be discussed later). As the exact shape of the two transformation functions is not known, they are represented as simple first-, second-, or third-order polynomials that are fit to the GCPs using least squares regression. The two first-order polynomials

$$r = a_0 + a_1x + a_2y$$

$$c = b_0 + b_1x + b_2y$$

control for translation in *x* and *y*, scale changes in *x* and *y*, skew, and rotation. Second- or third-order polynomials also control for warping that occurs with panoramic distortions. Rarely are polynomials of an order higher than three used. Though such high order polynomials may improve the accuracy of the transformation close to the GCPs, they can introduce severe errors and distortions in areas distant from GCPs (figure 5.13). Mathematically, the minimum number of GCPs needed to solve the polynomials is equal to the number of coefficients within an equation. For example, a minimum of three GCPs are needed for first-order, six for second-order, and ten for third-order polynomials (Richards 1986:52).

However, we are not only concerned with whether we can solve the polynomials; we are interested in the accuracy of the transformation. As the equations are solved using least-squares estimations, a rough rule of thumb is to select two to three times the number of GCPs as is needed mathematically. Not only is the number of GCPs important; their spatial distribution greatly affects the accuracy of the transformation. As we are working with a two-dimensional coordinate system, we should be careful to distribute GCPs fairly evenly over the image. We should avoid selecting GCPs that follow linear features, such as rivers, which may provide only one-dimensional information. We should also be careful to choose features that do not change location over time; we should thus avoid river meanders, sand banks, forest margins, and the like—especially if the image and map or GPS dates are substantially different.

The best control points are road intersections, edges of bridges or dams, corners of buildings, or other unambiguously identifiable permanent features. In industrialized countries with well-developed infrastructures, identifying sufficient numbers of GCPs is usually relatively simple. In wilderness areas and throughout much of the developing world, finding accurate maps of appropriate scale and identifying unambiguous GCPs can be extremely difficult, thus limiting the accuracy of geometric

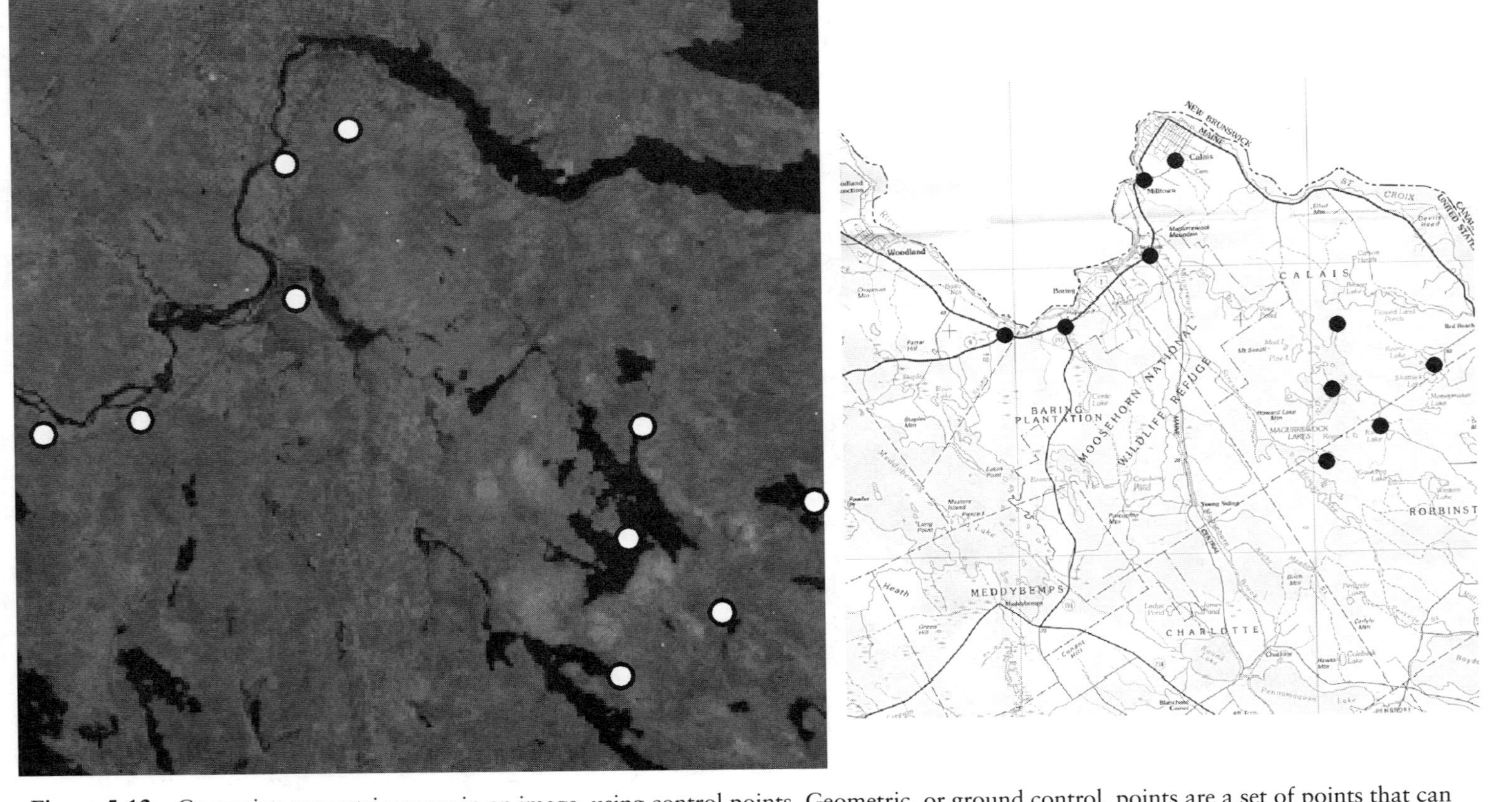

Figure 5.12 Correcting geometric errors in an image, using control points. Geometric, or ground control, points are a set of points that can be unambiguously identified in the image and on a map. Ideally, they are evenly dispersed throughout the image. The known locations of these points, determined from the map, are used in a polynomial regression to transform the location of each pixel within the raw image so that the resulting image is geographically accurate. An example of a geometric transformation using these points is shown in table 5.5.

Table 5.9
Using GCPs for Geometric Correction of an Image

	Original Image		Corrected Image				Coefficients for geocorrection functions	
GCP #	x	y	x'	y'	Errors		x	y
1	215	421	635632.80	5003764.50	1.115073	b0	619709.794825041	4979626.439487040
2	176	396	634484.40	5003057.00	1.088889	b1	0.032954357	0.000115384
3	183	309	634656.30	5000457.00	1.081162	b2	-0.000436361	0.033315167
4	87	231	631718.80	4998085.50	1.287153			
5	22	220	629632.80	4997807.00	2.654300		Total RMS error	1.83016
6	393	227	640984.40	4997971.50	0.756274			
7	502	174	644156.30	4996400.00	3.670993			
8	389	147	640828.10	4995514.50	1.350218			
9	443	104	642484.40	4994214.50	2.094682			
10	382	59	640593.80	4992928.50	0.967414			
	Column	Row	Easting	Northing				
			UTM					

NOTE: The transformations in this table pertain to the ten data points portrayed in figure 5.12.

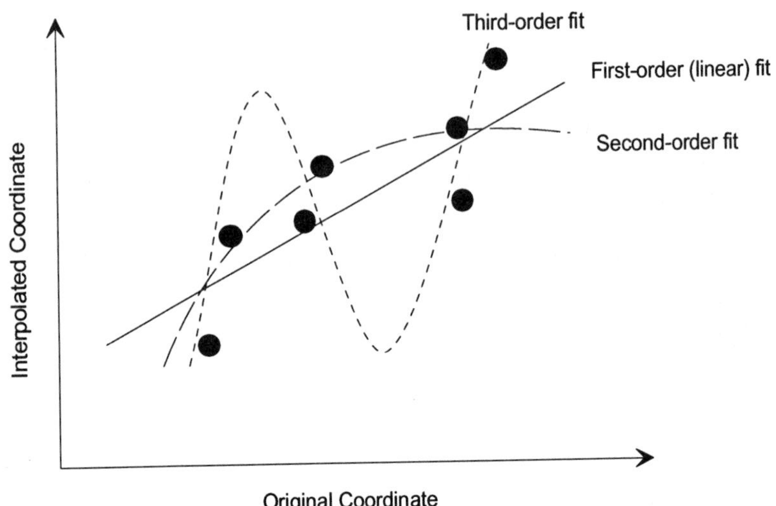

Figure 5.13 Impact of polynomial order on curve fitting and image rectification to a standard map projection. Note that the third-order polynomial curve passes closest to the greatest number of control points. To achieve this, however, sections of the third-order curve fall further from the control points than is true for any section of the lower order curves. As a result, a third-order polynomial transformation often produces very accurate results for areas of the image close to control points, but introduces severe geographical distortions in all other areas.

rectification (figure 5.14). When few GCPs are identifiable, more sophisticated rectification methods, such as those proposed by Friedman et al. (1983), may produce acceptable results.

Once GCPs have been selected, we can evaluate the accuracy of the polynomial transformation functions by computing the root mean square error (RMS) for each control point using

$$RMS_e = \sqrt{(\hat{r} - x)^2 + (\hat{c} - y)^2},$$

where x and y are the true geographic coordinates, and $\hat{r}$ and $\hat{c}$ are the transformed row and column locations within the raw image. Summing the individual RMS values provides an assessment of the total error in translating image pixel locations to true geographic coordinates. If the total error exceeds acceptable limits, then we can look at the individual GCP errors and delete from the analysis the worst points. In table 5.9, for

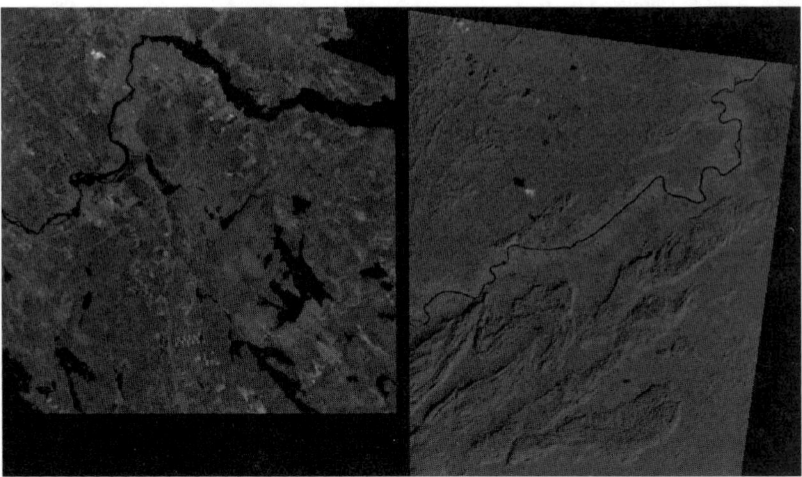

Figure 5.14 Images of landscapes rich and sparse in ground control points. The image of Calais, Maine, (*left*) is far richer than the image of eastern Honduras (*right*) in terrain features that can serve as ground control points. These ground control points can be matched with counterpart features on a standard reference map in order to correct for geometric error in the imagery.

example, the transformations in the chart would lead us first to drop GCP 7, then GCP 5. By deleting erroneous points, and recomputing the polynomial coefficients and RMS values, we can iteratively develop a more accurate transformation for the image.

Brightness Interpolation

Once we have developed an acceptable function to use in an image-to-map–based geometric correction, we can determine the "real" corner coordinates of a raw image under study and create a blank (all pixels are set initially to zero) corrected image large enough to contain the data from the raw image (figure 5.15).

We now must decide what brightness value to assign to each pixel in the corrected image. Even if the size of the pixels in the raw and corrected image are the same (this need not be the case), because the raw image was

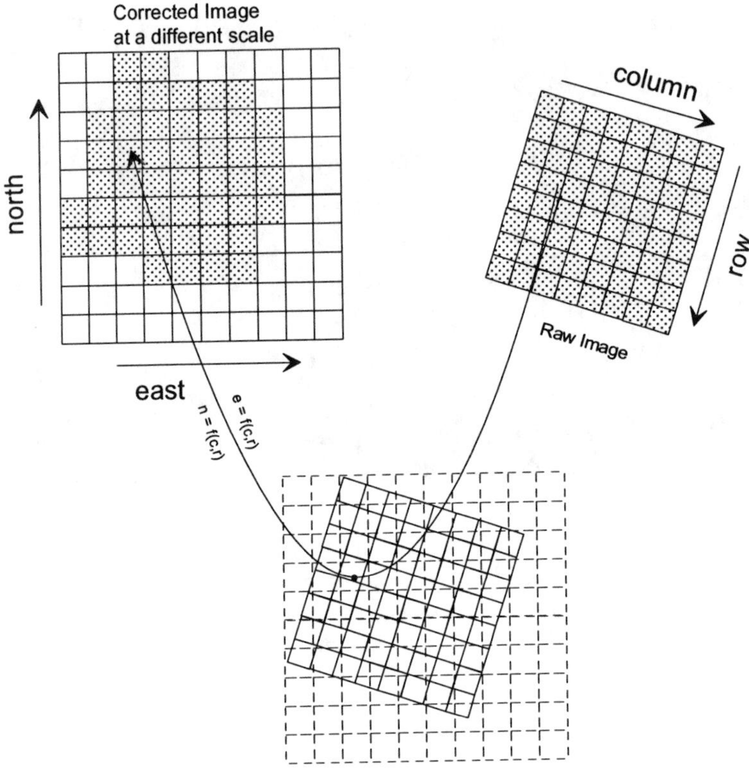

Figure 5.15 Mismatch in the overlay of raw and corrected image pixels. Each pixel in the raw image has row and column coordinates, whereas each pixel in the corrected image also has true geographical coordinates (i.e., row 4, column 5; latitude 41.3445, longitude 128.0348). During the transformation process the corrected image was warped to remove geographical errors in the raw image. Thus a pixel's location in the corrected image does not correspond directly with the location of any single pixel in the raw image. The brightness value to be assigned to the pixel in the corrected image must be interpolated from the neighborhood pixels of the corresponding location in the raw image.

distorted, it is not possible to overlay the raw image onto the corrected image such that each pixel in the corrected image is covered by a single pixel from the raw image. This is because the pixel locations (actually, the center points of each pixel) in the geometrically corrected image will be represented as real numbers rather than integers, and thus there is no direct one-to-one relationship between pixel locations in the raw image and in the corrected image. For example, pixel 4,3 (row, column) in the corrected image (representing, for example, the *x,y* coordinates 30.02, 19.03) translates to the raw image not as an integer pixel location but as location 3.8, 1.7. As the area immediately adjacent to this location falls among four raw image pixels (3,1; 3,2; 4,1; 4,2), we must interpolate a brightness value with which to fill the corresponding pixel in the corrected image.

The fastest method, and the one that is recommended because it retains raw brightness values, is *nearest-neighbor sampling* (figure 5.16). This method of interpolation is also very simple. One just selects the brightness value of the raw image pixel that is closest to the calculated pixel center. In our example we would use the brightness value of pixel 4,2.

Other techniques use the average brightness of several nearest pixels surrounding the calculated pixel center. The *bilinear interpolation method* uses four pixels; *cubic convolution* uses sixteen (Jensen 1986; Richards 1986). These more complex interpolations produce a less blocky effect than does nearest neighbor sampling, which often duplicates pixels. However, averaging of brightness values has a smoothing effect, resulting in loss of spectral detail that may have proved valuable for subsequent feature recognition and classification.

Simple Image-to-Image Coregistration

Once an image is geometrically corrected (rectified) it can be

- integrated with other digital map information, such as topography, soils, rainfall, etc. within a geographic information system spatial database;
- coregistered with an image of the same area recorded at an earlier or later date; and
- joined with other images to produce large area digital mosaics.

But if all that you need to do is register two images of the same area acquired at different dates (so as to allow pixel-by-pixel comparison), then it is not necessary to geocorrect the images to a standard map projection. Registration of multitemporal images can be achieved using a Sequential Similarity Detection Algorithm (SSDA) developed by Barnea and Silverman (1972). This method specifies one image as the "slave." Then,

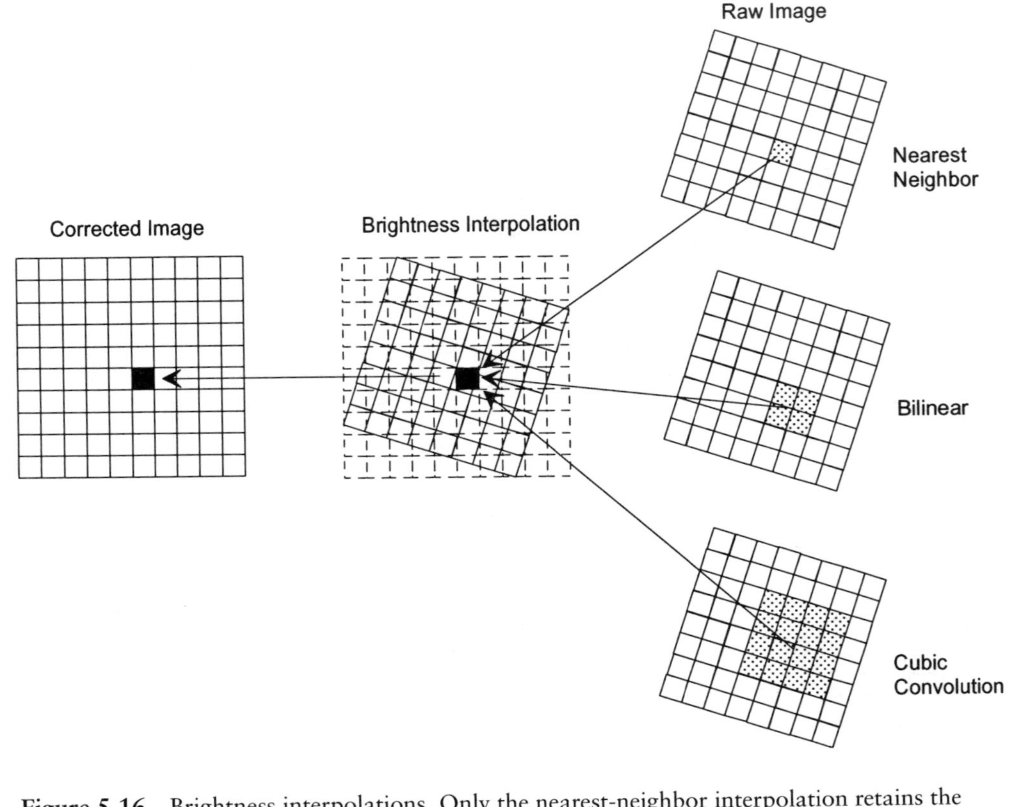

Figure 5.16 Brightness interpolations. Only the nearest-neighbor interpolation retains the brightness values of the raw image. The bilinear and the cubic convolution methods both represent averages and tend to smooth the resulting image.

using a series of control points selected from this image, the researcher attempts to locate the corresponding points within the "master" image. The premise of this technique is that the spatial orientation and brightness values of pixels in areas surrounding control points that have not changed in the time between acquisition of the two images should be highly correlated. If a window of pixels surrounding a control point is extracted from the slave image, and moved systematically over the master image in the approximately correct area, then the location that generates the smallest summed difference in pixel-by-pixel brightness values defines the matching control point as that lying under the center pixel in the moving window.

Clearly, this method depends on an absence of major spatial and spectral changes in the landscape in the time between acquisition of the two images. A more serious limitation of using this admittedly simple method is that though this technique allows two images to be coregistered and thus compared at the pixel level, both images retain the geometric distortions evident in the master image. As a result, distances between features, the area of features, and the relative positions of features will be inaccurate. The SSDA method only allows quantification of differences between images in relative rather than absolute terms. It is thus considerably less powerful than the polynomial least squares method already discussed.

6

Image Enhancement

Contrast Stretching for Scalar Image Enhancement
Linear Contrast Stretching
Nonlinear Contrast Stretching
Contrast Matching Between Images
Spatial Filtering for Scalar Image Enhancement
Low-Frequency Contrast Filtering (Image Smoothing)
High-Frequency Contrast Filtering (Edge Enhancement)
Vector Image Transformations
Vector Image Transformation by Addition, Subtraction, or Multiplication
Vector Image Transformation by Division
Transforming RGB Color Into Hue, Saturation, and Intensity (HSI)
Normalized Difference Vegetation Index (NDVI)
Kauth-Thomas Tasseled Cap
Principal Components Analysis

Successful interpretation of remote sensing imagery is almost always accomplished by visual characterization of features within a landscape. Nothing can take the place of the human eye and human judgment. A major goal, therefore, of digital image processing is to accentuate the visual cues that facilitate feature identification and thus interpretation of the imagery. This can be achieved by techniques that manipulate either single-band (scalar) images or multiband (vector) images.

Human vision is highly dependent on contrast (both as black-and-white tones and as color hues) for detecting and identifying features in a landscape. The major function of image enhancement, therefore, is to increase the contrast between the landscape background and the features of interest to the observer. Contrast enhancement used to assist visual interpretation of the image should not be applied prior to digital classification of the original data in the raw image, as the raw brightness values may be changed in unpredictable, nonlinear ways.

Single bands of an image, when first displayed, more often than not will appear either uniformly dark or uniformly light, with very little contrast across the image (figure 6.1 [upper]). There are two reasons for this lack of contrast. The first stems from an attribute of the detectors that produce the data for each band of an image. Detectors are designed to record the full range of brightness values exhibited by all landscape features on earth. For example, an ideal sensor must be sensitive to the very low brightness levels over oceans and deep lakes, and yet have a dynamic range that also can record the very high brightness levels reflected from snow and ice—without becoming saturated (or overexposed in photographic terms). As most landscapes on earth are not composed of features that cover the full dynamic range of brightness detectable by the sensor, the image will contain only a narrow range of brightness values and will appear dull or lacking in contrast.

The second factor responsible for low-contrast images is that many different landscape features reflect similar amounts of visible and infrared radiation, and will generate very similar (low contrast) brightness values. A hardwood forest, for example, that may contain very different species compositions may nevertheless appear homogeneous spectrally. Small ponds and adjacent areas of moisture-saturated soil may not be differentiable, and a village in Africa in which the huts are made of the same soil and grasses as the surrounding landscape may be impossible to distinguish from the background.

As discussed in chapter 3, the human visual system is incapable of perceiving more than 20–30 gray tones within a black-and-white photograph or image. The radiometric resolution of satellite sensors (MSS 128, TM 256, AVHRR 1024), however, greatly exceeds this threshold. Thus important features or trends within remote sensing imagery are often swamped in details that are beyond our ability to distinguish visually. A simple way to resolve this mismatch of technology and anatomy is *density slicing,* by which consecutive brightness values are clumped. Clumping reduces the number of brightness values (detail) within the image, but potentially increases the usable information content.

Density slicing can be used, for example, to transform the continuous gradient in brightness evident between shoreline and deep water into a discrete series of contours that often relate very closely to the bathymetry. Density slicing is most effective if the new brightness levels are displayed as colors (or graytones) that are graded along the spectrum rather than assigned arbitrarily.

Contrast Stretching for Scalar Image Enhancement

Improving the contrast of an image is achieved by stretching the narrow range of brightness values within the raw image in order to cover the full dynamic range of the computer's display. The brightness ranges of computer image processors are rated in bits, the same terminology that describes the radiometric resolution of satellite sensors. A 6 bit video processor can display 64 brightness levels (from 0–63), whereas an 8 bit system can display 256 tones simultaneously. The Landsat TM sensors have 8 bit resolution; therefore, an 8 bit video processor can display the full dynamic range of a single-band image recorded from this remote sensing system.

A computer system capable of displaying all the information (brightness values) within a three-band Landsat TM color composite would require a 24 bit (2^{8+8+8}) video processor. Computers with this capability can display more than 16 million colors simultaneously; these are the *true-color* systems. Memory demands for these kinds of systems can be daunting, as a computer must be able to store 24 bits of information in memory for each pixel in order to display 24 bit images. But as the price of computer systems continues to decline, *true-color* systems will become increasingly affordable. At present, 8 bit (Standard VGA on IBM compatible computers) and 16 bit (Super VGA) video processors are standard on most computer systems, and they are certainly sufficient for generating visually pleasing and interpretable black-and-white and even color composite images.

If we look at a brightness histogram (figure 6.2) for, say, band 1 of the Landsat TM image shown in figure 6.1, we can see that the contrast of this band does not make use of the full dynamic range of the display

(0–255). To enhance the contrast we can use two general techniques, linear stretching and nonlinear stretching.

Linear Contrast Stretching

To expand the contrast of a scalar (single-band) image in order to fill the dynamic range of a video display, we simply compute a new brightness value for each pixel ($BV_{stretched}$). We do this by inserting into the equation the raw image pixel brightness (BV_{raw}), the minimum (BV_{min}) and maximum (BV_{max}) brightness values of the raw image, and the dynamic range of the video display (V_{range}).

$$BV_{stretched} = \frac{BV_{raw} - BV_{min}}{BV_{max} - BV_{min}} \cdot V_{range} \ .$$

Schematically this *min-max* contrast stretch increases the spread of the brightness value histogram so that the raw minimum and maximum values are scaled for an 8 bit video display when V_{range} is set to 256. The minimum is scaled to zero and the maximum to 255. In many images the tails of the distribution may be exceedingly long or show a frequency spike at or near the minimum or maximum. When this is the case, a simple min-max stretch will only broaden the histogram slightly, thus barely improving the contrast. To remedy this, we can decide to select a BV_{min} and BV_{max} that are shifted inward from the true minimum and maximum values, thus excluding the outlier brightness values.

For example, in the histogram for band 1, there is a dominant frequency spike between 60 and 75, and a long low frequency tail at brightness values greater than 95. If we select 60 and 95 as the minimum and maximum, we can create a more effective *saturation* contrast stretch. If the histogram is approximately Gaussian, we can remove automatically the tails of the distribution by selecting the minimum and maximum at the bottom and top fifth or tenth percentiles (e.g., $BV_{\mathrm{min}} = BV_{5\%}$ and $BV_{\mathrm{max}} = BV_{95\%}$) resulting in a *percentage* contrast stretch. This method is displayed in the second row of figure 6.3. In both a saturation and a percentage stretch, data in the lower and upper tails of the histogram are lost, much as detail in a photograph would be, if some sections were under- and others over-exposed.

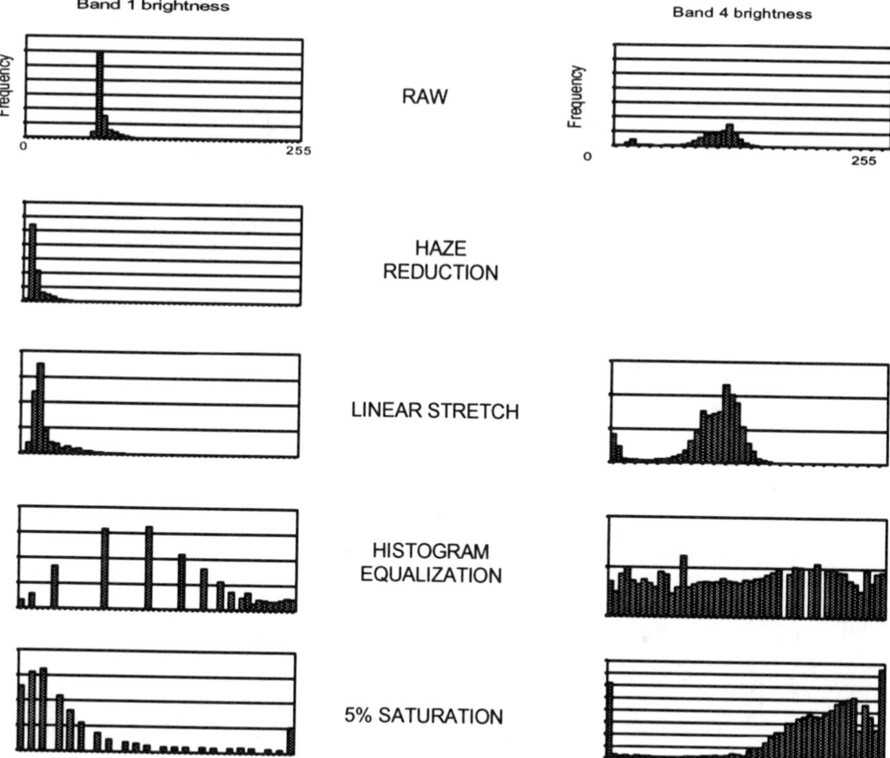

Figure 6.2 Brightness histograms derived from the raw image and from four different methods of scalar (single-band) stretching. Because the quality of a contrast enhancement is very site and observer dependent, several different methods should be undertaken, and the brightness histograms evaluated, before the best approach is selected.

Figure 6.1 Image enhancement by three different methods of contrast stretching. The effect of contrast stretching is shown here on the seven bands of the Landsat TM image window over Calais, Maine. *Upper:* linear min-max stretch. *Middle:* nonlinear histogram equalization stretch. *Lower:* 5% linear saturation stretch. For this particular image the histogram equalization stretch seems to best enhance the detail in bands 2 and 3; however, the 5% saturation stretch appears superior for all other bands. Note that stretching is a tool purely to improve the information visible within an image, and thus selecting the best stretch is very subjective.

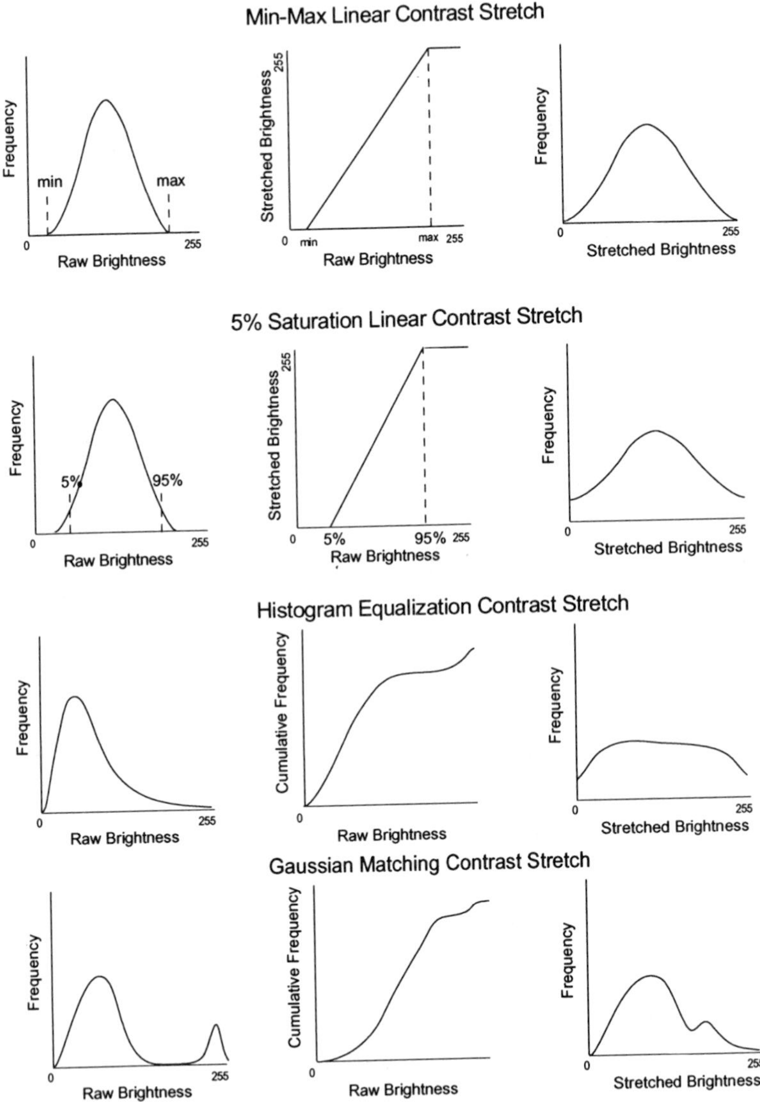

Figure 6.3 Four methods of contrast stretching for scalar image enhancement. The upper two methods are linear (that is, the brightness value in the stretched image is a simple arithmetic function of the raw brightness value). The lower two methods are nonlinear.

Nonlinear Contrast Stretching

If the raw histogram shows a very skewed, non-Gaussian distribution, then nonlinear contrast stretching may produce a more informative, visually appealing image, and should be tried if simple min-max or saturation stretching do not appear to work effectively. The two most common nonlinear stretch methods are *histogram equalization* and *Gaussian stretching.*

Histogram equalization Histogram equalization attempts to equalize the number of pixels present at each brightness level—that is, it attempts to make the height of each histogram bar the same. It attempts to do this across the full dynamic range of the video processor. In this way details should be visible at all brightness levels. Histogram equalization is thus a way to flatten the histogram such that each frequency bar contains as close to the same number of pixels as possible across the full dynamic range of the stretch. It is, we believe, a more effective method of scalar image enhancement than is Gaussian stretching. To create a uniform or equalized histogram from the raw brightness values, we first determine the total number of pixels within the raw image (N), and the probability of each brightness level containing pixels. We use the equation

$$\mathrm{p}_i = \frac{n_i}{N}$$

where (n_i is the number of pixels with brightness level i. We then select the dynamic range, R, for the stretched image. With N total pixels and R brightness levels (often referred to as bins), each bin in the stretched image should contain N/R pixels, and an equal probability ($p_i = 1/R$) of containing pixels. We can now assign raw image pixels to appropriate brightness level bins in the stretched image by examining the cumulative probability of each raw brightness value

$$P_j = \sum_{i=0}^{j} p_i,$$

and placing the pixels at this brightness (j) in the bin that has the closest cumulative probability. For example, suppose that brightness value 7 in the raw image had a cumulative probability of 0.66. Then with R=16 brightness levels in the stretched image, we should assign the raw pixels to stretched brightness level 11, as it has a cumulative probability of 0.6875 (P_{11} = 11/16).

Histogram equalization may not enhance the contrast of an image if the majority of pixels fall within only one or two brightness values (Richards 1986). This most often occurs in images that contain large homogeneous landscapes, such as lakes and snow fields. Under these conditions contrast may even be reduced, as many bins may remain empty, and low frequency brightness values are likely to be combined.

Gaussian stretching The Gaussian method of contrast stretching is accomplished in almost the same manner as that used for histogram equalization, but the height of the bars in the histogram match a normal distribution within, for example, two or three standard deviations around the mean (midpoint) of the dynamic range (R) chosen for the stretched image, with variance = 1. If we choose 16 brightness levels for the stretched image, then level 4 is the probability of containing a pixel that has a value of a normally distributed variable between -1 and -0.75 standard deviations from the mean.

Contrast Matching Between Images

In the previous two examples we mapped raw brightness levels onto histograms of predefined shape—either uniform in the case of histogram equalization or normally distributed in Gaussian stretching. It is also possible to use the histogram shape (cumulative probability histogram) of one image as the "standard," onto which the brightness levels of the other image can be mapped (figure 6.4). In this way the contrast of geographically adjacent images can be matched. By meshing the adjacent contrast-matched images together in a mosaic, a comparable and visually pleasing contrast across the whole image can be obtained.

Spatial Filtering for Scalar Image Enhancement

All contrast enhancing techniques described thus far focus only on the radiometric information contained within the image. But there is also a crucial spatial component. Our ability to detect and identify features within an image is, after all, largely a function of the contrasts between

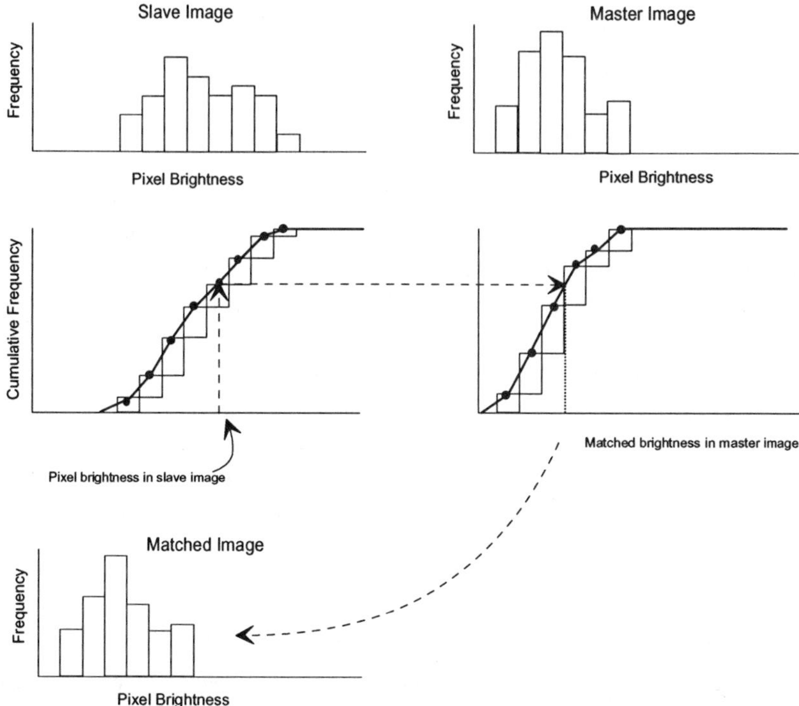

Figure 6.4 Contrast matching between images. Contrast matching is used prior to mosaicking two or more images together. By doing this we can remove the unnaturally abrupt change in tone that would otherwise be visible between images. Contrast matching uses the shape of the brightness histogram in one image (the master) to dictate the final shape of the brightness histogram for the other image (the slave) after transformation.

spatially adjacent features, or of a contiguous set of features along a gradient.

From a spatial perspective, we are more concerned with the change in brightness values that distinguish a feature of interest from the surrounds, rather than with differences in brightness levels for the image as a whole. If a landscape is rather homogeneous or shows a gradual contrast gradient, then the rate of change of brightness values as we follow a transect across the image will be slow. This is called *low-frequency detail.* If, on the other hand, brightness values change substantially over very short distances across the image, then the landscape exhibits *high-frequency detail.* Abrupt changes in brightness can occur at the edge of two adjoining features.

We can use this contrast frequency information to decide whether to accentuate or diminish the brightness between adjacent features in order to enhance the visual information available to an interpreter. To do this we use techniques that make use of the spatial elements within an image, rather than just the radiometric information. These spatial techniques all use moving templates (windows, masks, kernels) to alter the brightness value of the image pixel that lies under the center cell of the template. The alteration is performed according to an algebraic combination of the brightness values of the pixel's neighbors and the preset values contained within all template cells. The templates act as user-definable filters that selectively block (deemphasize) or enhance certain spatial contrast attributes of the image. Not surprisingly, these enhancements are often referred to as image filtering.

Low-Frequency Contrast Filtering (Image Smoothing)

Low-frequency, or low-pass, filtering is used to remove or deemphasize high-frequency (abrupt changes in brightness) information within an image that may be present as a result of random noise errors. Low-frequency filtering may also be used for image smoothing, if landscape heterogeneity is high and landscape patch size is small, relative to pixel size. An excess of landscape heterogeneity is a common problem in images of highly diverse and patchy environments, such as tropical moist forests and suburban fringes.

The two most commonly applied kinds of spatial filters for image smoothing transform the brightness value of a pixel according to either (1) the mean brightness value of neighboring pixels or (2) the median brightness value.

Mean brightness filtering Mean-value filtering is accomplished by passing an odd-sided, square template (3×3, 5×5, etc.), with cell values of 1, over the raw image, and transforming the brightness of the image pixel that lies under the central cell of the template. The brightness of this central target pixel is transformed into the sum of the products of the

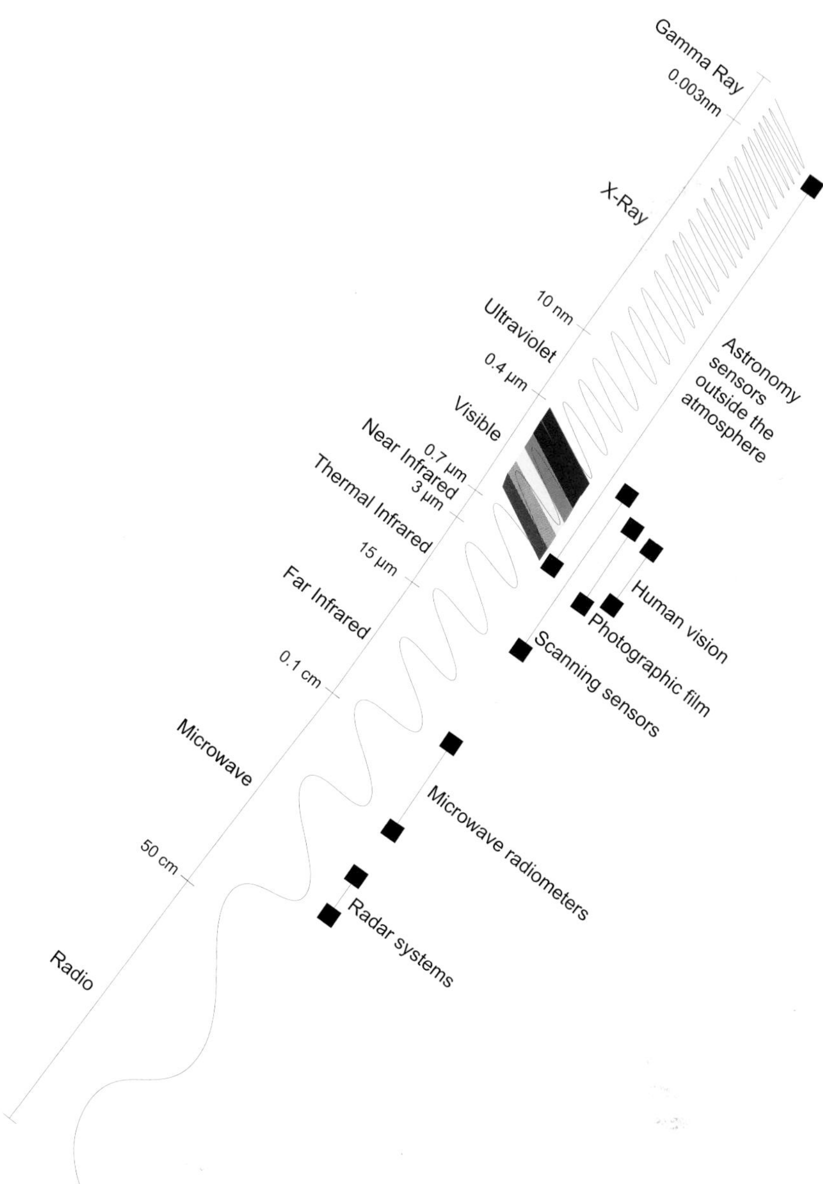

Plate 1 Electromagnetic radiation and the sensitivity range of remote sensing systems. (nm = nanometers = 1 x 10^{-9}m; {gm}m = micrometers = 1 x 10^{-6}m; cm = centimeters = 1 x 10^{-2}m)

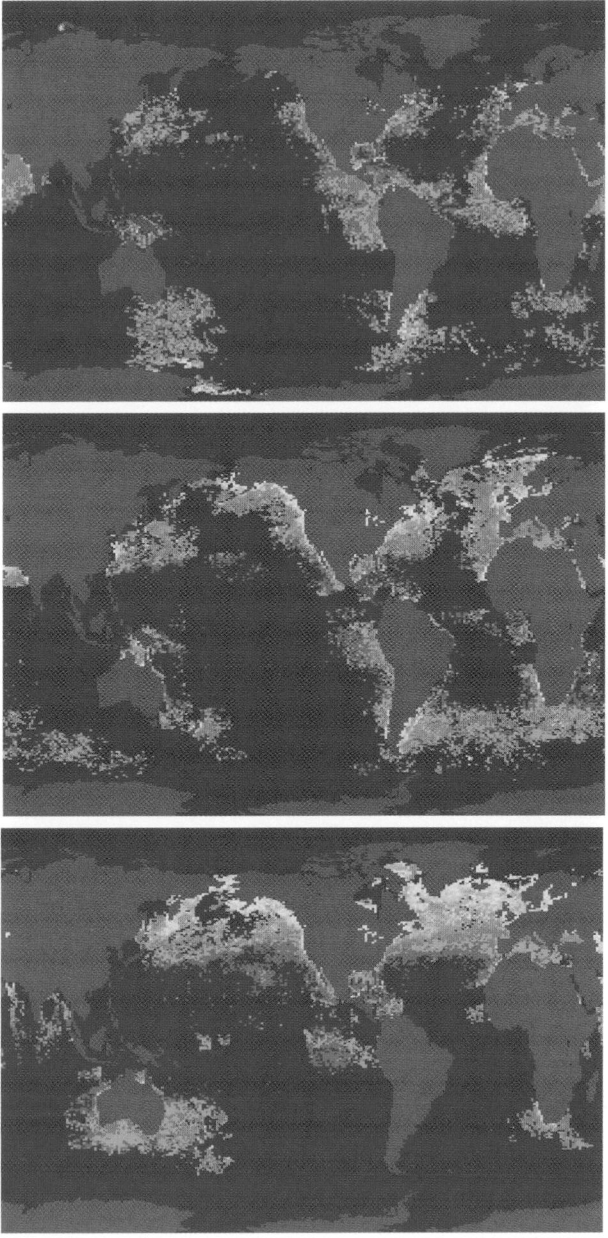

Plate 3 Global marine phytoplankton maps from CZCS data. Seasonal changes in global marine phytoplankton distribution and abundance are evident here. Green and yellow represent the quantity of chlorophyll in the water. The images were recorded by the CZCS (Coastal Zone Color Scanner) in January (*top*), May (*middle*), and September (*bottom*).

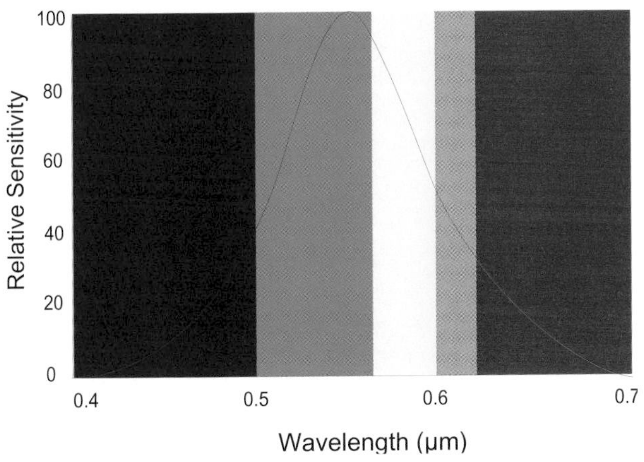

Plate 2 Spectrum of visible radiation and the sensitivity of human vision.

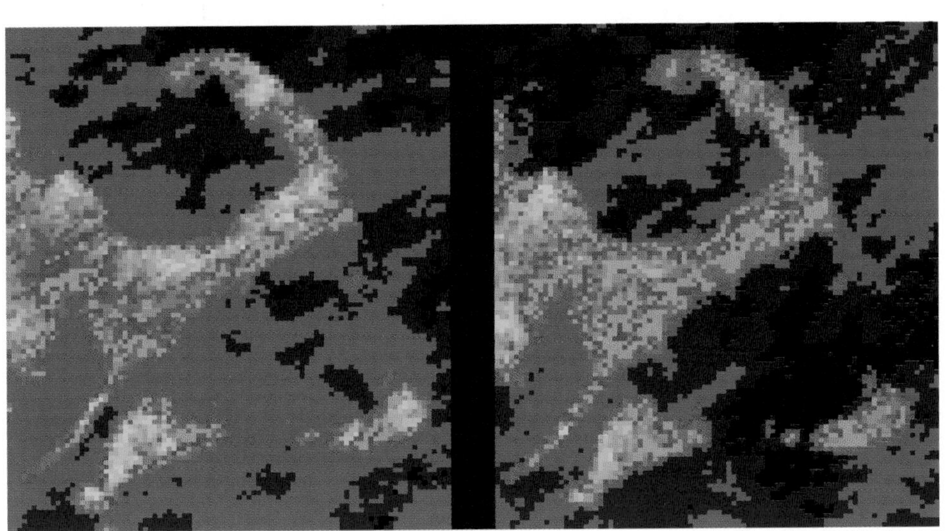

Plate 4 Effects of Hurricane Bob. Hurricane Bob blew up the east coast of the United States on 20 August 1991, passing over Cape Cod and eastern Massachusetts. The storm lasted less than four hours, but its intense winds blew down many trees, and salt spray caused extensive leaf damage in the region. Shown here are two AVHRR composited NDVI images that represent the green leaf biomass present within the landscape. *Left:* A composite of daily images taken within the ten days preceding the hurricane. *Right:* A composite of daily images taken during the ten days following the storm. The after-storm image (note the increase in brown areas) shows a marked decrease in chlorophyll along the track of the storm.

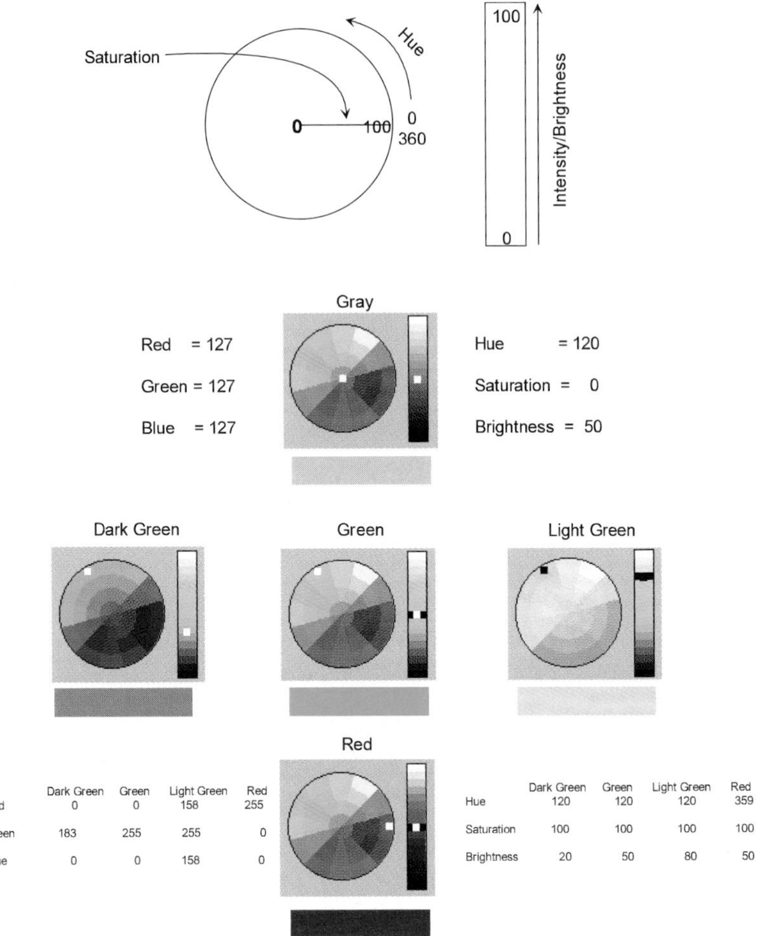

Plate 5 Comparison of the RGB and HSI systems for displaying colored images. A computer display screen (monitor) is composed of a matrix of dots, each composed of phosphors that emit red, green, or blue (RGB) light when excited by an electron beam. This system mimics the human eye, where the color-sensitive cells, the cones, are of three types—each sensitive to red, green, or blue light. Colors on a computer monitor are formed by exciting the RGB phosphors at different intensities. If the relative RGB intensity is equal, a series of gray tones is produced that ranges from black (where the RGB phosphors are not excited) to white (where they are at maximum intensity). When only one phosphor type is excited, a pure (100%) saturated hue is produced. Mixed colors are obtained when the RGB intensities are unbalanced (unsaturated). Different tones (brightnesses) of a pure or mixed color are obtained by maintaining the RGB ratios but increasing or decreasing the intensity. The HSI system is used by televisions and videocameras to display colors. In this system *hue* changes with the angle of an arc around a circle. *Saturation* (how pastel the color is) varies from zero at the center of the circle to 100 at the perimeter. The lightness or darkness of the color (e.g., light blue or dark blue) is determined by changing the brightness (*intensity*).

Plate 6 Examples of three-band RGB color composite images. Selecting the best band combinations to display as three-color (blue, green, red) composites depends on the dominant landscape features within the imagery and the features that the researcher would most like to distinguish from the background. From the far left, the five image windows shown here display band combinations that exhibited the three highest optimum index factor values (band combinations 1,3,4; 1,4,5; and 3,4,5) and the two lowest values (1,2,6 and 2,3,6). The OIF values are based on the total variance and intercorrelation of the three-band combinations (see table 5.4). Note that although the band 1,3,4 combination had the highest OIF value, and combination 1,2,6 the second lowest, the latter because the colors are more natural looking appears to convey more information, at least visually.

Plate 7 Assessing image quality. Random pixel errors and atmospheric haze are evident in this true-color (bands 1, 2, 3) composite of Landsat TM imagery over a region in Maine. Haze is visible most clearly in the top right corner of the image. Random errors are scattered throughout the image and are visible as individual oddly colored pixels that stand in contrast to the like-colored background.

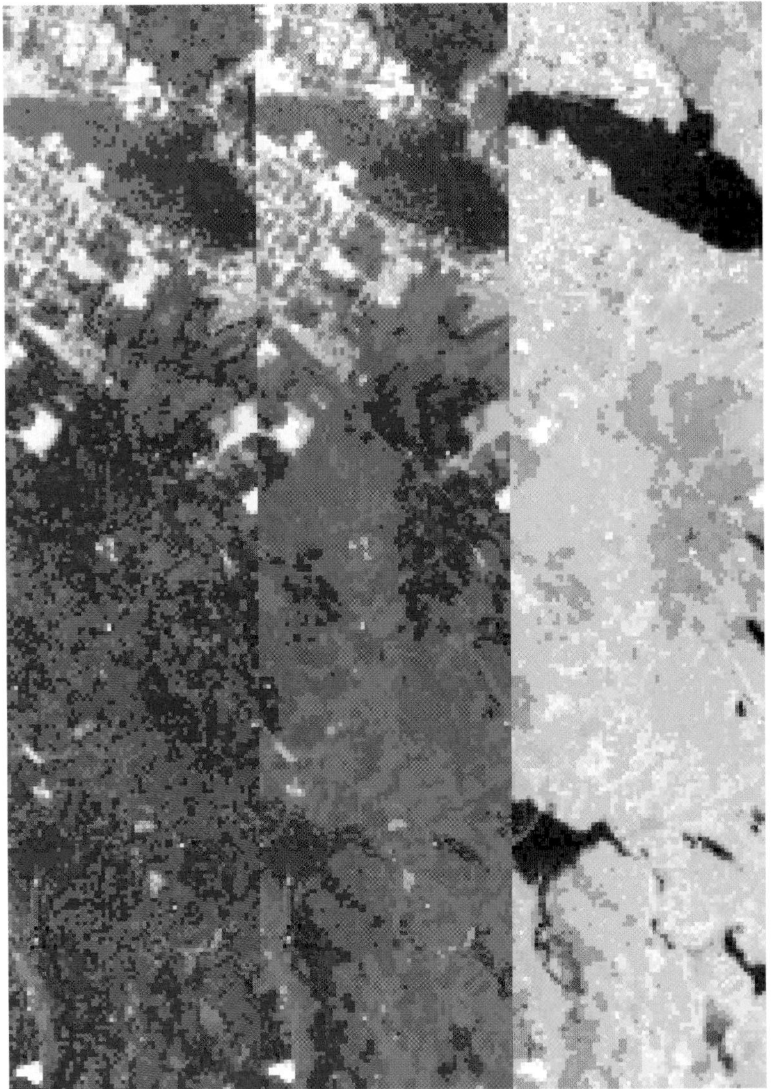

Plate 8 Color composite images created from PCA components. These images were created by displaying the first three PCA components (shown in figure 6.16) as RGB color composites. By varying the color assigned to each component, different features are emphasized in the resulting composite image. The 123 (1 = red, 2 = green, 3 = blue) RGB combination (*left*) appears to highlight urban areas most effectively. The 321 RGB combination (*middle*) shows improved detail within open water areas. The 312 RGB combination (*right*) emphasizes differences within vegetated areas. This disparity is not surprising because in each case the dominant color is generated primarily from component 1, which contains the highest contrast (the most information) of any component.

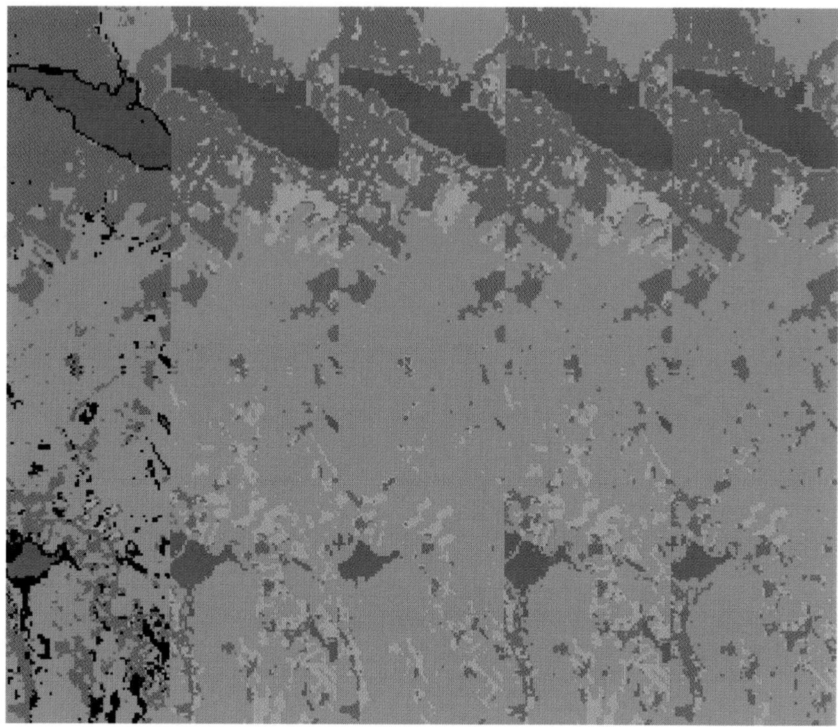

Plate 9 Comparison of three supervised classification methods. Based on the information obtained from plotting the brightness profiles of each training set (see figure 7.12), Landsat TM bands 2, 4, and 5 were selected for the supervised classifications displayed in the first three images starting at the left. The classifiers used were parallelepiped (*far left*), minimum distance to centroid (*second left*), and maximum likelihood (*middle*). In each image, blue = water, gray = urban, green = forest, yellow = cleared, and black = unclassified. Notice that only the parallelepiped classifier (*far left*) resulted in some pixels remaining unclassified. Comparison of all three classifiers shows that the minimum distance classifier labeled many more pixels as cleared than did the other two classifiers. Without verifying the accuracy of the classification, estimates of deforestation would vary wildly depending on the classifier used. To explore the data further, minimum distance (*second right*) and maximum likelihood (*far right*) classifications using only bands 4 and 5 were completed. Visual comparison of the classifications using three bands versus two bands show very few differences. This suggests that band 2 added little to the discriminating power of bands 4 and 5 combined.

template cells and underlying image pixels, then divided by the number of cells in the template, according to the equation

$$target_{r,c} = INTEGER\left[\left(\sum_{x=r-(n/2-1)}^{r+(n/2-1)} \sum_{y=c-(n/2-1)}^{c+(n/2-1)} raw_{x,y} template_{x-(r-n/2),y-(c-n/2)}\right)/n^2\right]$$

where r is the row location, c is the column location of the target pixel in the raw image, and n is the edge dimension of the template. Figure 6.5 presents a visual example of how a mean brightness transformation is accomplished. Though mean filtering is effective in removing random noise, like any moving average statistic it will deemphasize abrupt changes in brightness, thus muting the true edges at the boundaries of landscape features. The blurring that results from mean filtering will increase with the size of the template used (figure 6.6). But it can be somewhat reduced by center weighting of the values in the template.

Median brightness filtering An alternative to mean filtering (and one that is less likely to result in severe image blurring) uses the median of a target and its neighboring pixels. Median filtering has two advantages over the neighborhood average brought about by mean filtering: (1) the median of an odd-sided square template will always equal the brightness value of one of the neighborhood pixels; and (2) the median is much less affected than the mean by extreme data values.

Figure 6.7 portrays how these two methods can lead to very different results. In this example, which uses a 3×3 template, the neighborhood pixel values were 2, 1, 26, 3, 2, 9, 4, 2, and 10. The median (the central value when the pixel brightness values are ranked in ascending or descending order) is 4, compared to a mean of 6.55. The mean (7 when rounded) is not a brightness value present in the raw image, and it is larger then six of the nine neighborhood pixels. The median, conversely, appears more representative of the raw data.

While removing possibly spurious extreme values, the median retains much of the detail in the raw image (figure 6.8). The major drawback of using median filtering is that it is computationally much more intensive than is mean filtering. Mather (1987) suggests, however, that this can be minimized by assigning the median to be the brightness level that shows

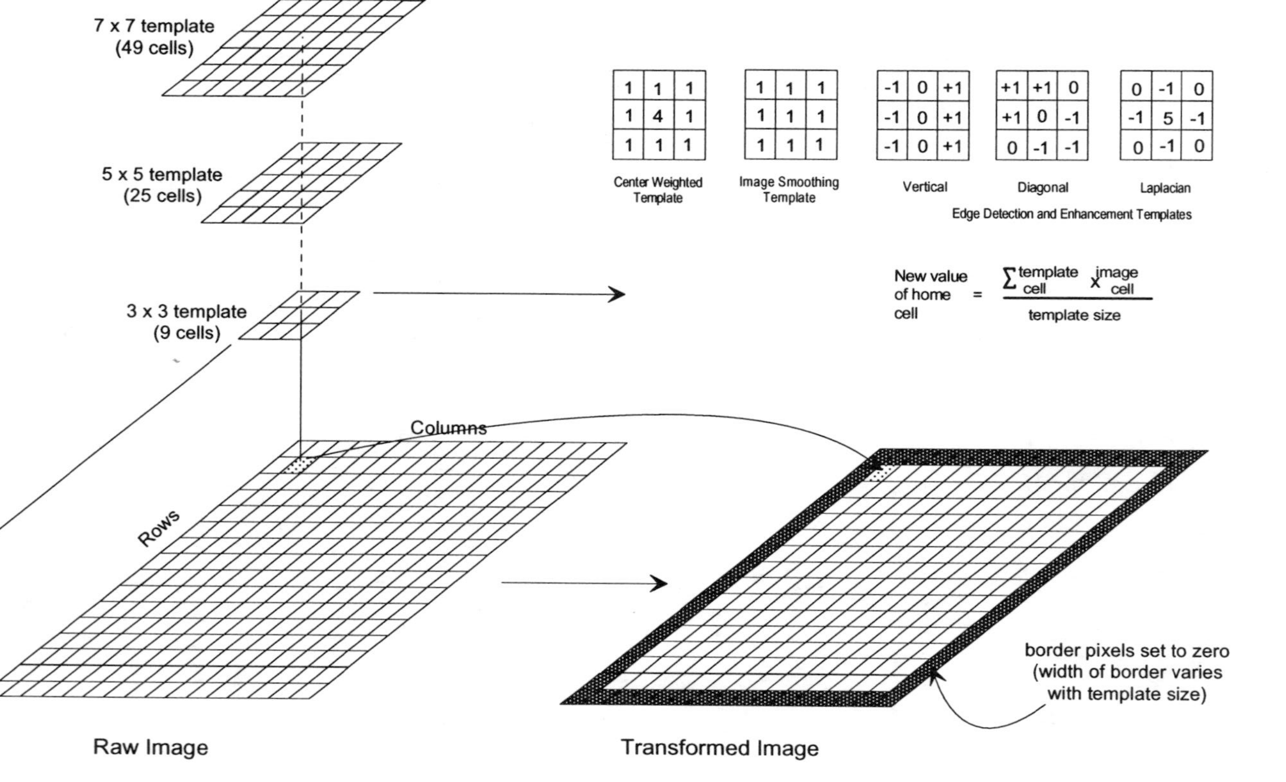

Figure 6.5 Templates for spatial filtering of an image. For example, the image smoothing template replaces the home cell brightness with the mean brightness of all cells covered by the template (i.e., each cell covered by the template is multiplied by the number in the overlying template cell, in this case 1, and the average of these nine numbers replaces the brightness value of the home cell in the image). The center-weighted template is also a smoothing averaging filter, but by multiplying the home cell by

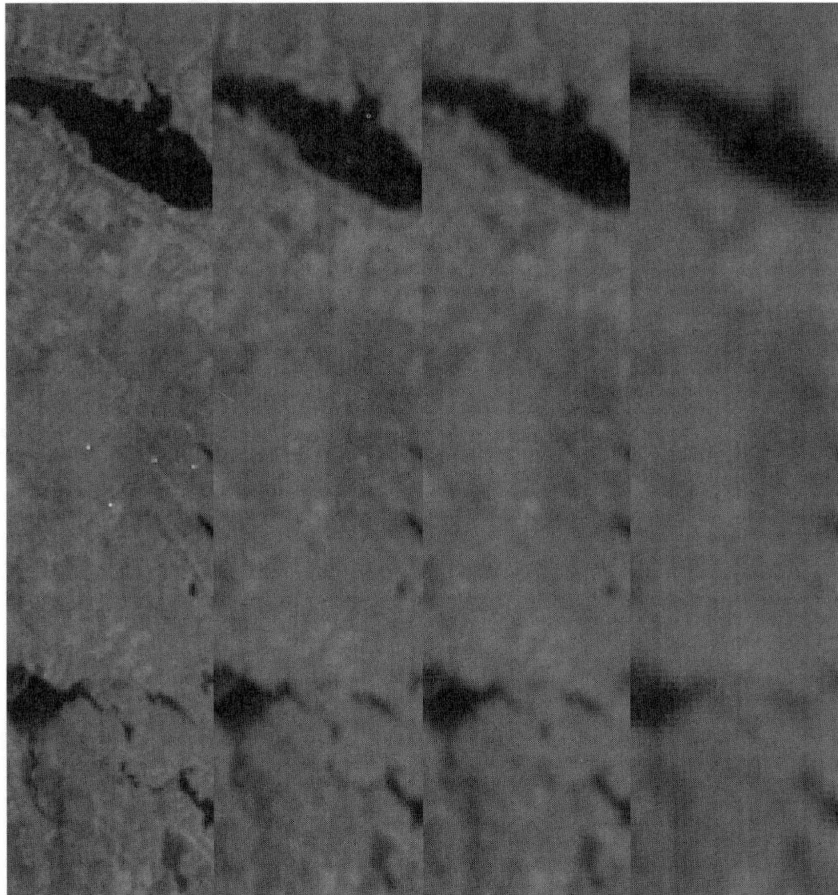

Figure 6.6 The mean brightness method for image smoothing. The three images shown here demonstrate the progressive blurring associated with increasing the template dimensions. *Left:* 3×3. *Middle:* 5×5. *Right:* 7×7.

a cumulative frequency greater or equal to $(n/2)+1$, where n is the edge dimension of the template.

High-Frequency Contrast Filtering (Edge Enhancement)

As satellite imagery from all sensor systems is constructed by compositing the brightness of all landscape features within a pixel, detail or contrast

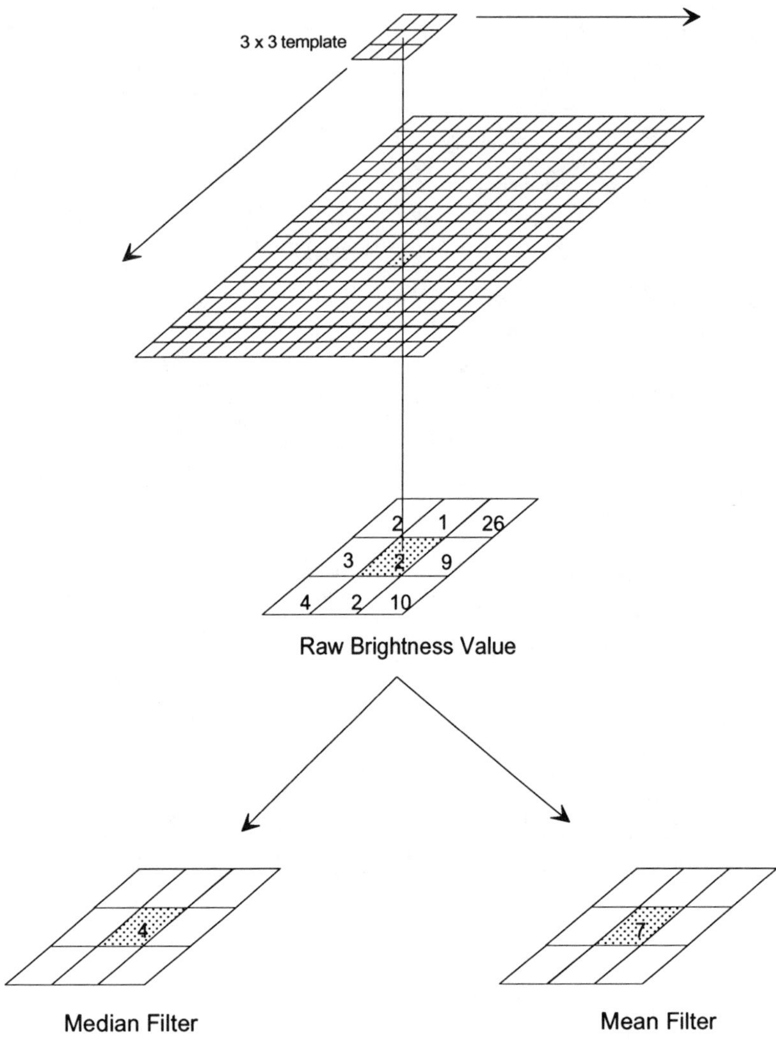

Figure 6.7 Comparison of median and mean spatial filtering methods for random noise removal. In this example a raw pixel brightness of 2 would be transformed to a brightness value of 7, using the mean brightness of the template covering pixels. The transformation would yield a brightness of 4 if the median were used. The median is more representative of the brightness of the home cell's neighbors, whereas the mean is not.

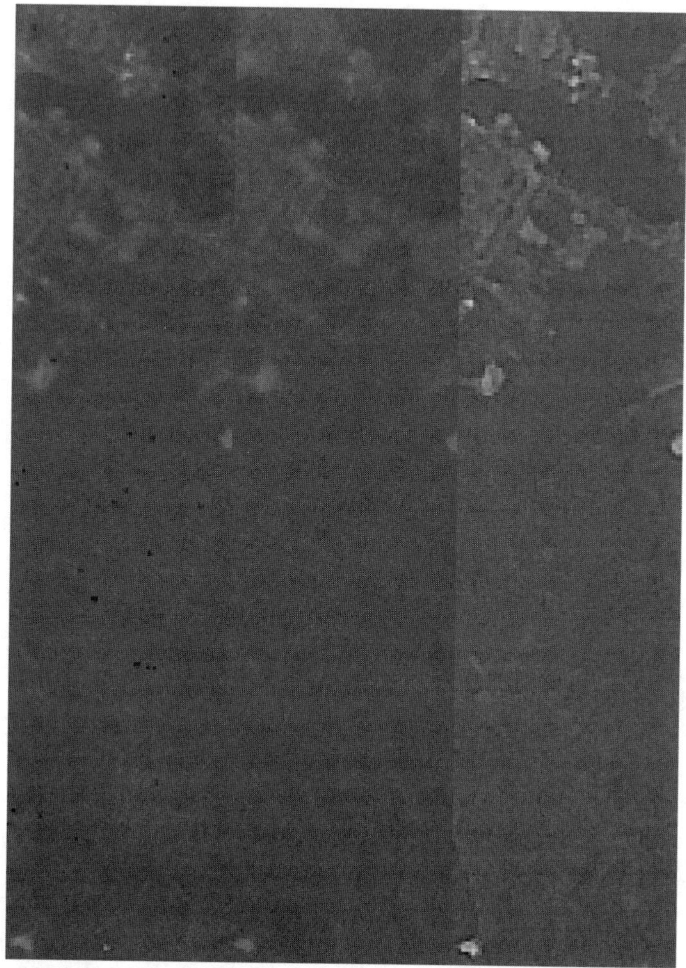

Figure 6.8 Random noise removal and edge enhancement. *Left:* A window of Landsat TM band 1, where random detector errors give the image a peppered appearance (the isolated black pixels are the errors). *Middle:* The same image window with random errors removed using a median spatial filter. Notice how the peppered effect has gone, but the sharpness of the image has been reduced (smoothed). *Right:* By applying an edge sharpening spatial filter to the middle image, much of the lost visual detail has been restored.

is lost in the process. If we can enhance the edges of features, however, we can increase the perceived detail and potentially our ability to interpret the image. Emphasizing all the high-frequency components of an image is achieved by either: (1) removing the low-frequency components and adding the detected edges back into the raw image, or (2) applying a Laplacian template to the raw image.

Accentuating the high-frequency detail by subtracting the low-frequency information is accomplished very simply, using the equation

$$High\ frequency_{r,c} = 2(Raw_{r,c}) - Low\ frequency_{r,c}$$

where the brightness value of a pixel in the low-frequency filtered image is subtracted from twice the pixel brightness value in the raw image (figure 6.8). High-frequency filters can also enhance linear features aligned in a specific direction. This is particularly useful in extracting surface features that may indicate ground water sources, or other hydrological characteristics of an area. For example, a north-south template will emphasize all vertically oriented linear features or edges. Rotating the north-south template by 90 degrees will emphasize all east-west (horizontally) oriented features. Other orientations can be detected by setting the template diagonal values to 0 and the off-diagonals to -1.

A Laplacian template applied to a raw image emphasizes the contrast of edges within the image regardless of their orientation. That is, a Laplacian enhancement in mathematical terms removes the blurring that occurs as a result of imaging through the atmosphere and the use of optical lenses, so that a sharp point or edge no longer appears out of focus. This template is equivalent to creating an image that has positive and negative pixel values, representing the peaks and hollows of contrast gradients, and then adding this edge-detected image back into the raw image. The result is an enhancement that increases the detail visible within the image, thus improving our ability to detect and interpret features.

Although edge detection and enhancement filters are very effective in emphasizing the shape of features, thus improving visual interpretation of imagery, digital shape recognition is only in its infancy in remote sensing analysis.

Vector Image Transformations

Thus far we have discussed methods of scalar image enhancement—that is, methods for increasing the visibility of useful information *within a single spectral band* by either image smoothing or edge enhancement. Image smoothing is useful for weeding out random errors; edge enhancement is useful for better drawing out features from the background. Vector image transformation techniques go the next step, by moving beyond a single band.

Whereas a pixel in a scalar image is represented by a single (scalar) value, a pixel within a vector image is represented by several (a vector) of brightness values. For example, a pixel on a computer screen is composed of red, green, and blue intensity values. A multispectral remote sensing image such as that generated by SPOT HRV can be considered as a vector image composed of three scalar images representing green, red, and IR reflectance respectively.

Vector image transformations can be used to emphasize features or create new ones by combining information within *multiple spectral bands*. If two images are coregistered (that is, when each pixel overlays exactly its corresponding pixel in the other image or images), then we can add, subtract, multiply, and divide corresponding pixels within the images. Although all four algebraic operations are possible, image addition by itself has little practical value.

Vector Image Transformation by Addition, Subtraction, or Multiplication

Addition of all bands within a multispectral image will yield a single image that represents the total reflected energy from the landscape (figure 6.9). The primary value of image addition is in GIS operations, such as recombining two masked images and as the last step in a more valuable transformation, such as image multiplication.

Subtraction is useful for detecting temporal change in a particular terrain imaged at two different times. Subtraction of one image from the other will generate an image for which the brightness values of each pixel represent the degree of change that has occurred between the image acquisition times (figure 6.10). Image differencing and landscape change detection will be discussed in more detail later.

Multiplication is most often used to mask out sections of an image. Such masking may be desirable, for example, when isolating haze-covered areas before performing a *haze removal* on the image. In this third form of vector transformation, the raw image is multiplied with a Boolean

Figure 6.9 Additive image composed of all seven Landsat TM bands. Band 6 was resampled to 30m from its raw spatial resolution of 120m prior to creating the additive image. This image contains the total brightness of the landscape at all wavebands. Additive images are of little interpretive value, as we no longer can use our knowledge of the spectral response of features in different wavebands to characterize areas according to their brightness.

Figure 6.10 Using image subtraction to detect temporal change. This difference image was derived from two AVHRR NDVI images of Cape Cod, one taken before and the other after Hurricane Bob. The very light pixels within the image indicate areas where leaf biomass was most reduced by the storm. Notice how the light areas are clustered along the coast, and particularly on the eastern edge of the peninsula and Nantucket Island.

image. The Boolean (or binary) image is composed of just two values: zeros for pixels to be excised from the raw image and ones for pixels to be retained. A Boolean image can be created by thresholding an image; brightness values less than the threshold are set to zero, those greater are set to 1. When the Boolean image is multiplied with the raw image on a pixel-by-pixel basis, the resulting image retains the original brightness values in the areas of interest (set to 1 in the Boolean image), but all other areas are blank (i.e., have a brightness of zero). This technique can also be used to differentially stretch specific sections of an image that have very different reflectance characteristics (land versus water), and then recombine them by image addition to recreate the full scene.

Vector Image Transformation by Division

Image division is the most useful and commonly used arithmetic operation undertaken on pairs of bands within a multispectral image. With multispectral imagery such as Landsat TM, there are $(n^2-n)/2$ spectral band combinations (where n = number of spectral bands) that could be used to generate a *ratio image.* (This assumes, of course, that the thermal band 6 was resampled to 30m resolution, to make its resolution compatible with the other bands.) However, ratio images are not usually created out of all the possible Landsat bands. Rather, a ratio image most often is created by combining just the visible red and near-infrared bands (Landsat MSS bands 5 and 7; Landsat TM bands 3 and 4). This specific combination of bands has been found to generate an image that discriminates very effectively among soil, water, and vegetation, and can help to remove topographic shadows that can confuse feature identification and classification.

Image acquisition time of most earth resources satellites is set away from solar noon to avoid specular (mirrorlike) reflection problems. Consequently, the combination of sunlight angle and topography often results in irradiance differences across the landscape. Therefore, similar features will exhibit different brightness values when situated on the illuminated and shadow sides of landscapes with even moderate relief. In figure 6.11, for example, the brightness values for red and near-IR bands over an area of coniferous forest differ depending on sun angle and topography. But, interestingly, the ratio of these bands is almost identical. Thus, near-IR/red ratio images are able, where topographic shadows are not saturated (black), to reduce the effects of topography. Transformation by division reveals the land-cover patterns more clearly, allowing comparable landscape features to be associated more accurately.

A second interesting attribute of near-IR/red ratio images stems from differences in spectral response curves of vegetation, soil, and water between the visible red and near-IR portions of the spectrum. Figure 6.12 is an example of a division transformation (ratioing) of Landsat TM bands 3 (red) and 4 (IR). The ratio images 3/4 (second to right) and 4/3 (far right) are the inverse of one another—notice that water is white in the 3/4 ratio image and black in the 4/3 ratio image. Both ratio images, when compared to the raw bands 3 (far left) and 4 (second to left), enhance our ability to distinguish between urban (nonvegetated) features

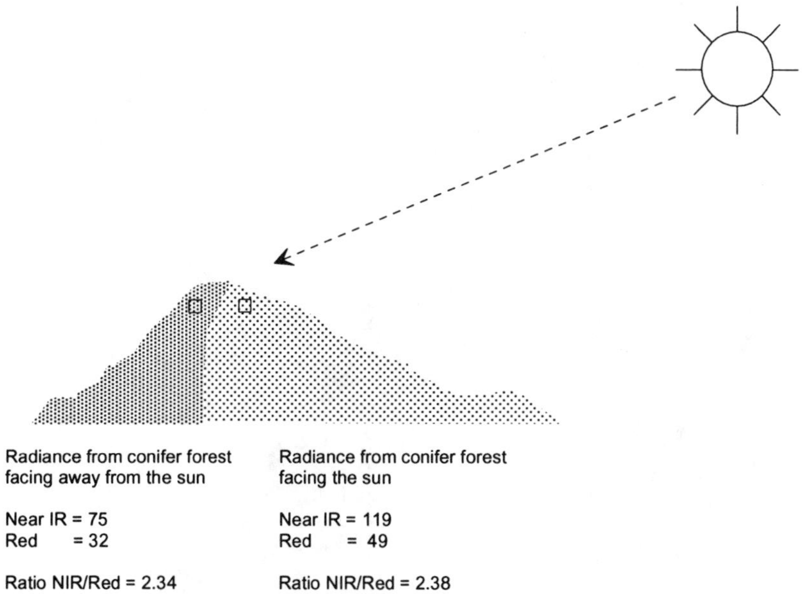

Figure 6.11 Topographic shading and how to correct for it. In this example of coniferous forest, a division transformation of the two spectral bands yields very nearly the same ratio value (2.34 and 2.38), whether the forest is on the sunny or shaded side of a hill.

and the surrounding fields and forests. This is particularly evident in the interface between urban and nonurban landscapes at the top of each image window.

Figure 6.13 shows graphically how the different spectral signatures of water, soils, and vegetation result in very different Landsat MSS band 7 (near-IR) to band 5 (red) ratios. Though band 5 is able to distinguish soils from vegetation, and band 7 can discriminate water and vegetation, only the 7/5 ratio is able to separate all three features from one another. By ratioing the two bands, we can emphasize the slope of the spectral reflectance curve between 0.8–1.1 μm and 0.6–0.7 μm. In the resulting ratio image, large positive values represent healthy vegetation, because healthy vegetation exhibits a characteristic and dramatic increase in reflectance between the visible red and near-IR regions of the spectrum. Similarly, in very general terms, values close to 1 indicate soils; those less than 1 represent water.

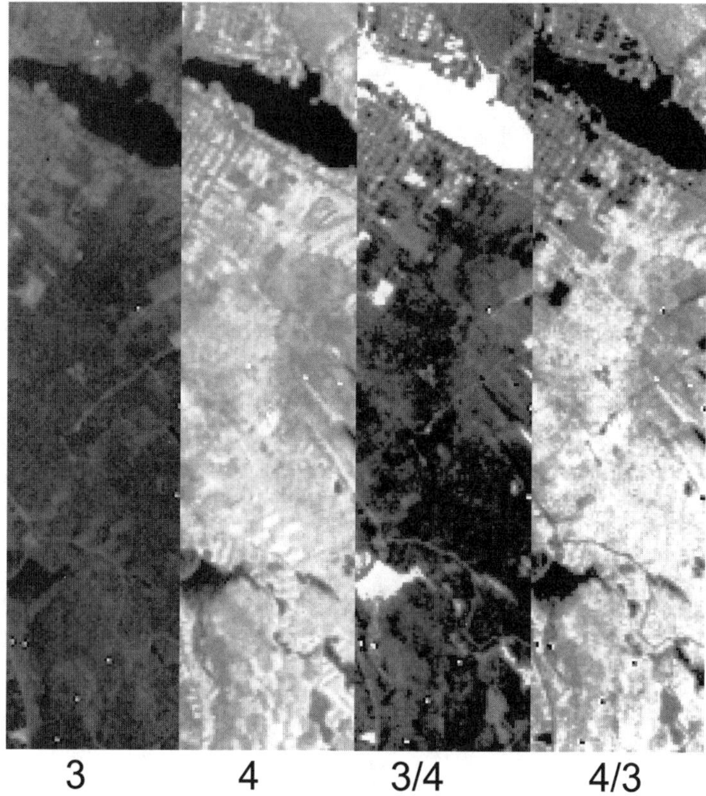

Figure 6.12 Using vector division (image ratioing) to differentiate water, soils, and vegetation. The two ratios (3/4 [second to right] and 4/3 [far right]) are the reciprocals of one another. Notice that water is white in the 3/4 ratio image and black in the 4/3 ratio image. Both ratio images, when compared to the raw bands 3 (*far left*) and 4 (*second to left*), enhance our ability to distinguish between urban (nonvegetated) features and the surrounding fields and forests. This is particularly evident in the interface between urban and nonurban landscapes at the top of each image window.

A large number of ratio images can be computed for multispectral imagery. Choosing the most useful ratio image often involves trial and error, guided by an understanding of the spectral response of features of interest relative to background features. Landsat MSS band 6/5 ratios have been found useful for regional plant phenological and biomass change studies in the continental United States. Band 4/5 ratio images for Landsat MSS

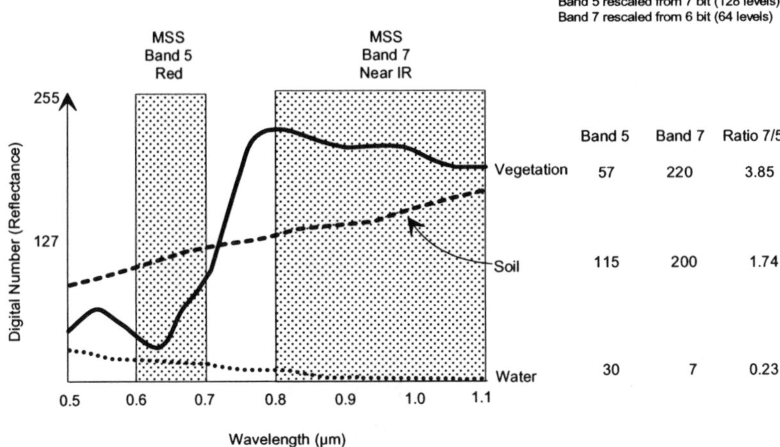

Figure 6.13 Schematic of vector division for differentiating terrain features. The schematic shows graphically how the different spectral signatures of water, soils, and vegetation result in very different Landsat MSS band 7 (near-IR) to band 5 (red) ratios. Though band 5 is able to distinguish soils from vegetation, and band 7 can discriminate water from vegetation, only the 7/5 ratio is able to discriminate all three features from one another.

tend to minimize vegetation variation across a landscape, making it easier to discriminate soils and soil condition.

As ratio images range typically in value from zero to 5 or 6, they should be stretched before being displayed, so as to enhance the contrast. Alternatively by computing

$$Ratio_{r,c} = 162.34 \left(\arctan \frac{near\text{-}IR_{r,c}}{red_{r,c}} \right)$$

the resulting ratio can range from 0–255. To avoid division by zero when computing ratios, add 1 to any pixels in the raw images that have zero brightness.

Transforming RGB Color into Hue, Saturation, and Intensity (HSI)

As we discussed in the previous chapter, image color can be represented either by combining the primary colors (red, green, and blue in different amounts) or by defining each pixel according to the three components of color: (1) *hue* (dominant wavelength); (2) *saturation* (purity of the hue, which indicates the amount of white mixed in with the color; as white increases, the color becomes more pastel); and (3) *intensity* (brightness or dullness of the hue). Hue, saturation, and intensity are together abbreviated as HSI. Transforming RGB components of a composite image into HSI components provides a tool for combining images from different sources or with different resolutions. This transformation is also useful for generating a 4-band composite image (tables 6.1 and 6.2).

To generate a very visually pleasing image that incorporates the spectral resolution of a 3-band Landsat TM color composite with the spatial resolution of a SPOT 10m panchromatic image, we do the following:

1. Resample the Landsat TM images to a spatial resolution of 10m.
2. Coregister the Landsat TM images with the SPOT image.
3. Convert the Landsat TM images from RGB to HSI.
4. Rescale the SPOT panchromatic image to fractions with a range of 0–1.
5. Multiply the intensity by the rescaled SPOT image.
6. Convert the HSI data back to RGB for display.

The same process can be used to generate any 4-band image using either four multispectral bands or three multispectral bands plus a radar image or digital elevation data.

Normalized Difference Vegetation Index (NDVI)

Simple ratio images are very effective in discriminating vegetation, soil, and water. If, however, we are interested in examining how vegetation density changes over time, comparison of simple ratio images may be made difficult by changes in weather conditions and solar irradiance. The normalized difference vegetation index (NDVI) was developed to overcome this problem. This index is computed by dividing the difference of the near-IR and visible red bands (MSS bands 7 and 5, or TM bands 4 and 3) by their sum. Global daily NDVI data are computed from AVHRR bands 2 and 1 at a 4km resolution. These computations have proved ex-

Table 6.1
Conversion Algorithm to Transform from RGB to HSI

```
(rescale image data to 0–1 before running the algorithm)
intensity = max (r,g,b)
mn = min (r,g,b)
range = intensity − mn
IF intensity ≠ 0
   saturation = range/intensity ELSE saturation = 0
ENDIF
IF saturation = 0
   r′ = (intensity − r)/range
   g′ = (intensity − g)/range
   b′ = (intensity − b)/range
   IF intensity = r
      IF mn = g THEN hue = 5 + b′ ELSE hue = 1 − g′
      ENDIF
   ENDIF
   IF intensity = g
      IF mn = b THEN hue = 1 + r′ ELSE hue = 3 − b′
      ENDIF
   ENDIF
   IF mn = r THEN hue = 3 + g′ ELSE hue = 5 − r′
      ENDIF
   hue = hue * 60
ELSE
   hue = −1 (undefined)
ENDIF
(rescale hue, saturation, and intensity to 0–255 before displaying)
```

SOURCE: Adapted from Mather 1987:230–231.

Table 6.2
Conversion Algorithm to Transform from HSI to RGB

```
IF hue = 360 THEN hue = 0 ENDIF
hue = hue/60
h′ = int(hue)
h″ = hue − h′
x = intensity * (1 − saturation)
y = intensity * (1 − (h″ * saturation))
z = intensity * (1 − saturation * (1 − h″))
SWITCH(h′)
     CASE 0:        r=intensity      g=z              b=x
     CASE 1:        r=y              g=intensity      b=x
     CASE 2:        r=x              g=intensity      b=z
     CASE 3:        r=x              g=y              b=intensity
     CASE 4:        r=z              g=x              b=intensity
     CASE 5:        r=intensity      g=x              b=y
```

SOURCE: Adapted from Mather 1987:230–231.

tremely useful for continental monitoring of changes in vegetation pattern and density over time, as will be discussed later.

Kauth-Thomas Tasseled Cap

Kauth and Thomas (1976) developed a vegetation index that employs four rather than two spectral bands. Using Landsat MSS imagery, they were able to show that vegetation at different growth stages occupies a graphed space shaped like a tasseled cap, when plotted in three orthogonal dimensions. The three directions are named *brightness* (soils), *greenness* (growing vegetation), and *yellowness* (senescing vegetation). A fourth orthogonal dimension named *non-such* accounts for the variance in brightness values not attributable to soils and vegetation. When land-

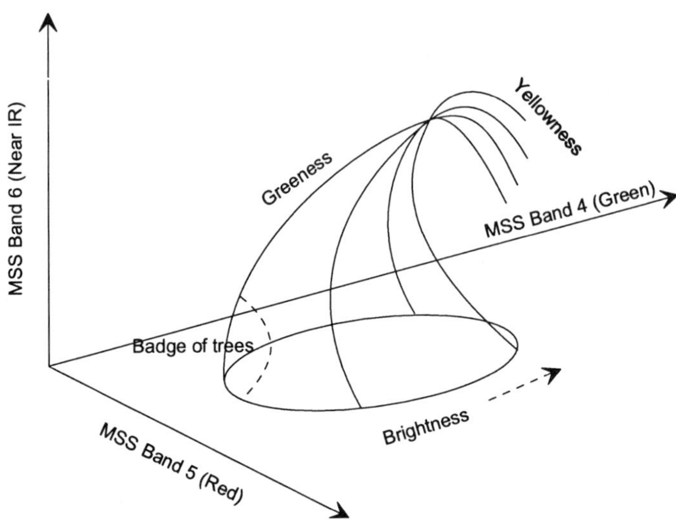

Figure 6.14 Kauth-Thomas 3-D "tasseled cap." When the brightness of pixels in Landsat MSS bands 4, 5, and 6 are plotted in three dimensions, vegetation pixels clump together in a shape that resembles a tasseled cap. The tassel is formed by senescing (yellowing) vegetation. Pixels dominated by trees tend to clump in a "badge" at the front of the cap. As it is not possible to represent graphically four dimensions, the non-such Kauth-Thomas dimension is not shown here. After Richards 1986:144.

scape reflectance data within an image are plotted in three dimensions, pixels that represent forest or woodlands cluster in the front of the cap as a badge of trees. Figure 6.14 depicts in three dimensions the tasseled cap feature that stimulated development of this transformation.

Coefficients for the transformation are calculated by determining the characteristics of a line with the best fit through a set of soil classes extracted from the imagery. The second "greenness," third "yellowness" and fourth "non-such" axes are then computed using a Gram-Schmidt orthogonalization procedure. To create the Kauth-Thomas tasseled cap components of brightness, greenness, and yellowness, the brightness values in the raw MSS image are combined in these four equations:

$$\text{Brightness} = 0.433\ \text{Blue} + 0.632\ \text{Green} + 0.586\ \text{Red} + 0.264\ \text{Near-IR}$$

$$\text{Greenness} = -.290\ \text{Blue} - .562\ \text{Green} + 0.600\ \text{Red} + 0.491\ \text{Near-IR}$$

$$\text{Yellowness} = -.829\ \text{Blue} + 0.522\ \text{Green} - .039\ \text{Red} + 0.194\ \text{Near-IR}$$

$$\text{Non-such} = 0.223\ \text{Blue} + 0.012\ \text{Green} - .543\ \text{Red} + 0.810\ \text{Near-IR}$$

Looking at the size and sign of the coefficients in each equation shows the following: Soil (brightness) is a weighted sum of all bands. Vegetation biomass (greenness) is ostensibly the difference between the long red and near-IR wavebands and the short green–blue bands. Senescing or stressed vegetation (yellowness) is largely the contrast between visible blue and green bands.

The Kauth-Thomas transformation was designed as an agricultural crop investigation tool, and so its utility for biodiversity conservation will depend on the question being addressed. Like much remote sensing analysis, trial and error are essential for extracting the most useful information from the imagery.

Principal Components Analysis

The Kauth-Thomas transformation is just one of several techniques that create orthogonal axes from often highly intercorrelated bands within multispectral imagery. The most common of these techniques is *principal components analysis* (PCA). The PCA form of vector transformation not only removes the correlation (redundancy) between bands; it also effectively reduces the number of bands within an image without losing overall information content. PCA is a *decorrelation contrast stretch* that often increases the visual information (features) within an image by combining the information in multiple bands for simultaneous display. PCA compresses the variance in a multispectral image (using linear transformations) such that the first three bands in a PCA image contain over 95% of the information (variance) contained in the original *n*-band image. This is the equivalent of changing your viewpoint of an object (swarm of data) so that it casts the largest shadow.

The PCA transformation works on imagery of any number of dimensions. Four-band Landsat MSS, 7-band Landsat TM, and 224-band AVIRIS imagery can all be transformed using PCA. To explain how the transformation works, we will use only two dimensions. If we plot the visible green and near-IR bands against one another (figure 6.15), we can see that the swarm of data aligns itself rather tightly along the diagonal that separates the two main axes. In other words, these two bands show a very strong positive correlation, where an increase in reflectance in one band is matched by a linear increase in the other. High interband correlation means that if a composite image were created, little contrast would be evident, as data are clustered along the main diagonal rather than distributed evenly in the space between the two axes. If we stretched the data within each band using the techniques for linear contrast stretching (discussed earlier in this chapter), the only thing we would accomplish is an elongation of the data along the main diagonal—which would not improve the overall contrast of the image.

However, as most of the data are aligned along the main diagonal, we could reduce the dimensionality of the image without losing information by replacing the two axes with the diagonal along which all the data are aligned. This is exactly what PCA does. It defines the principal axes of the image, first by locating the axis that maximizes the variability (contrast) across all dimensions, then by adding perpendicular (orthogonal) axes that account monotonically for decreasing variance within the image.

In the two-dimensional example portrayed in figure 6.15, principal component (axis) 1 is located along the main diagonal and PC2 is located at right angles to it. If we then complete a simple linear stretch of each principal component, the data swarm will almost fill the space between the axes, and we will have dramatically increased the contrast and thus

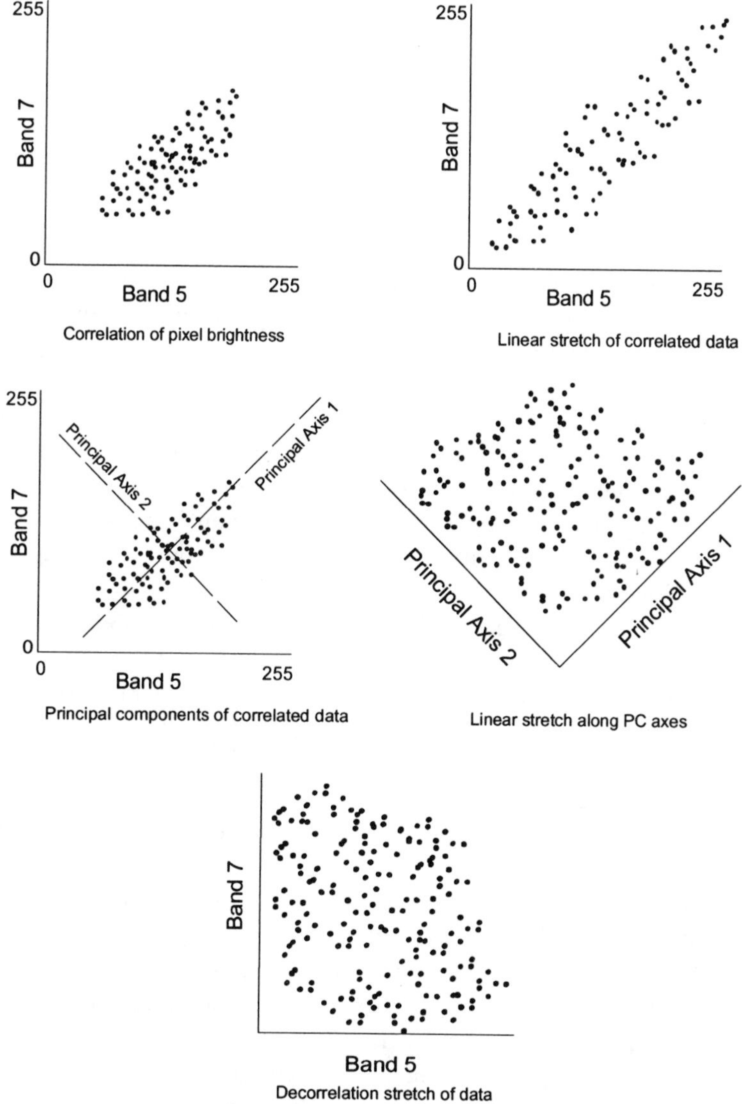

Figure 6.15 A 2-D example of a PCA (principal components analysis) transformation. In this example, when pixel brightness values for Landsat TM bands 5 and 7 are plotted in two dimensions, the data aggregate along an invisible line, indicating that the data within the two bands are highly correlated. Stretching the brightness values merely elongates the swarm of data along the correlation axis. But PCA rotates the original axes so that the principal axis is oriented to represent the greatest variance in the data. Stretching the data within the new axes spreads the brightness values more evenly, resulting in a final image with much more visual contrast.

the amount of information visible within the image. For two-dimensional problems it is reasonably easy to identify the first component by examining a bivariate plot of the data. Multidimensional imagery cannot be examined geometrically, and so principal components must be determined algebraically.

The PCA transformation uses a variance-covariance matrix to calculate a coefficient (eigenvectors) matrix, which when applied to the brightness values of raw image pixels yields a PCA (decorrelation) transformation. This transformation generates the same number of components as there are bands within the raw imagery, although each component contains progressively less variance (i.e., less contrast or information). For example, the first three components computed from Landsat TM imagery will typically contain over 95% of the variance within all spectral bands. Thus a color composite made from these three principal components will contain almost all the information available within the original multiband imagery.

Figure 6.16 shows images of the seven components (numbered 1 to 7 from the far left) generated from a PCA transformation of the seven spectral bands in the Landsat TM image over Calais, Maine. Notice that from left to right the amount of detail in each image declines. Principal components 1 and 2 (far left and second to left) contain most of the information contained within the seven original spectral bands. Components 5, 6, and 7 are almost uniform in tone, and appear to contain almost no useful information.

Plate 8 combines the first three components into RGB color composite images. By varying which color is assigned to each component, different features are emphasized in the resulting composite image. The component 123 RGB combination (left) appears to highlight urban areas most effectively, whereas the 321 combination (middle) shows improved detail within open water areas, and the 312 combination (right) emphasizes differences within vegetated areas. This is not surprising because in each case the dominant color is generated primarily from component 1, which contains the highest contrast (the most information) of any component.

One disadvantage of PCA is that it is an automatic process that is not guided to emphasize the features we are interested in and want to distinguish. Thus, although high-order principal components will contain less variance than will the low-order components, the information that the

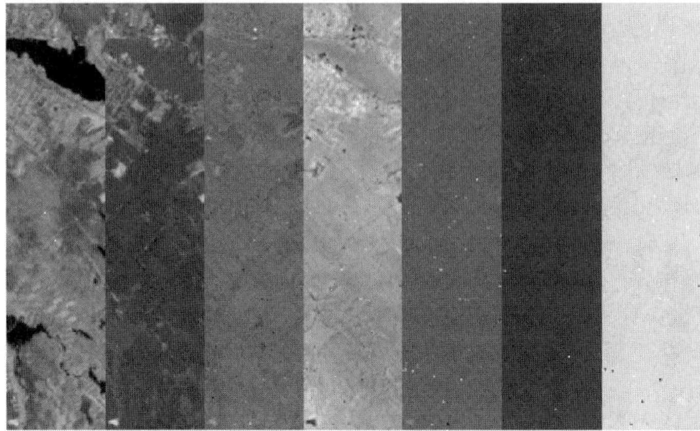

Figure 6.16 Result of a PCA transformation of the Calais, Maine, Landsat TM image window. The seven component images are numbered 1 to 7 from the far left. Notice that from left to right the amount of detail in each image declines (i.e., the contrast or brightness variance declines). Principal components 1 and 2 (far left and second to left) contain most of the information captured within the seven original spectral bands. Components 5, 6, and 7 are almost uniform in tone, and appear to contain almost no useful information.

low-order components contain may nevertheless be useful. Before the low-order components are discarded, therefore, they should be examined individually to determine if one or another does indeed distinguish valuable features. For example, in figure 6.16 components 3 and 4 clearly provide less information than does either component 1 or 2. However, component 4 does seem to emphasize features that border water bodies, which may be useful for some applications. Applying the transpose of the matrix to the PCA values will regenerate the raw-image brightness values for each band (adding some integer rounding error in the process).

The major disadvantage of using PCA is that the brightness values of each component do not relate directly to the reflectance characteristics of earth surfaces. As a result, we cannot use our knowledge of landscape reflectance characteristics to help detect and interpret features visible within PCA images. Feature identification within PCA images is very dependent then on their location, orientation, shape, and juxtaposition—all of which can only be made sense of if we have ancillary information, such as aerial photographs, land use maps, or field experiences to help guide the interpretation.

7

Feature Identification and Classification

- *Unsupervised and Supervised Methods*
  - *Unsupervised Classification*
  - *Supervised Classification*
- *Selecting a Land-Cover Classification Scheme*
- *Preparing Training Sets*
- *Setting Up a Field Survey*
  - *Choosing a Time for Data Collection*
  - *Preliminary Image Analysis to Guide the Field Survey*
  - *Determining the Composition, Size, and Number of Field Sites*
  - *Selecting an Appropriate Sample of Field Sites*
- *Executing a Field Survey*
  - *What to Measure and How to Measure It*
  - *Recording the Location of Each Field Site*
- *Fine-Tuning the Training Sets*
- *Selecting and Using a Classifier*
  - *Parallelepiped Classification*
  - *Minimum Distance to Centroid Classification*
  - *Maximum Likelihood Classification*
- *Improving the Accuracy of the Classification*
  - *Determining the Accuracy of the Classification*
  - *Improving Classification Accuracy*
- *Detecting and Monitoring Temporal Change in a Landscape*
  - *Coregistration of Change Detection Images*
  - *Strategies for Detecting Change*
- *Generating Maps of the Analysis Results*

In the previous chapter we described strategies and techniques that emphasize and accentuate the spectral and spatial features contained within an image. In this chapter we explain how to label these features so that they represent real landscapes that can be identified on the ground—landscapes that, when combined, characterize the area bounded by an image. Our discussion in the previous chapter of principal components analysis (PCA) for image interpretation revealed some of the problems in detecting and identifying features within remote sensing imagery. So we begin this chapter with a survey of the kinds of ancillary information, and the image processing tools, essential for relating features visible within the imagery to corresponding discrete objects and landscapes observed on the ground.

Image analysis and classification methods and the order of implementation of individual components will vary to some extent depending on the problem to be solved. Figure 7.1 depicts a fairly standard scheme for conducting a digital remote sensing image classification. Notice that following image correction and enhancement, the classification process moves into either a supervised or an unsupervised approach. We begin this chapter accordingly.

Unsupervised and Supervised Methods

The image enhancement techniques described in the previous chapter are designed to emphasize the contrasts within the image (edge enhancement) or to diminish the contrasts posed by shading and random pixel error (image smoothing). These preliminary steps are undertaken in the hope that the manipulated spectral information represented as color and tonal differences will aid our visual ability to detect and separate discrete spectral features within the image. Unsupervised and supervised classification are the digital analogs of this visual separation of spectral features. Both strategies use the spectral characteristics of pixels and the researcher's knowledge of the landscapes present within the image to assign each pixel to a unique and recognizable land-cover type. The two strategies differ from one another primarily in the order that spectral information and ground-based information are used to classify the imagery.

Unsupervised classification first statistically clumps spectral features in the imagery into discrete classes. Then the researcher, using maps and field-based knowledge, assigns each class as unambiguously as possible

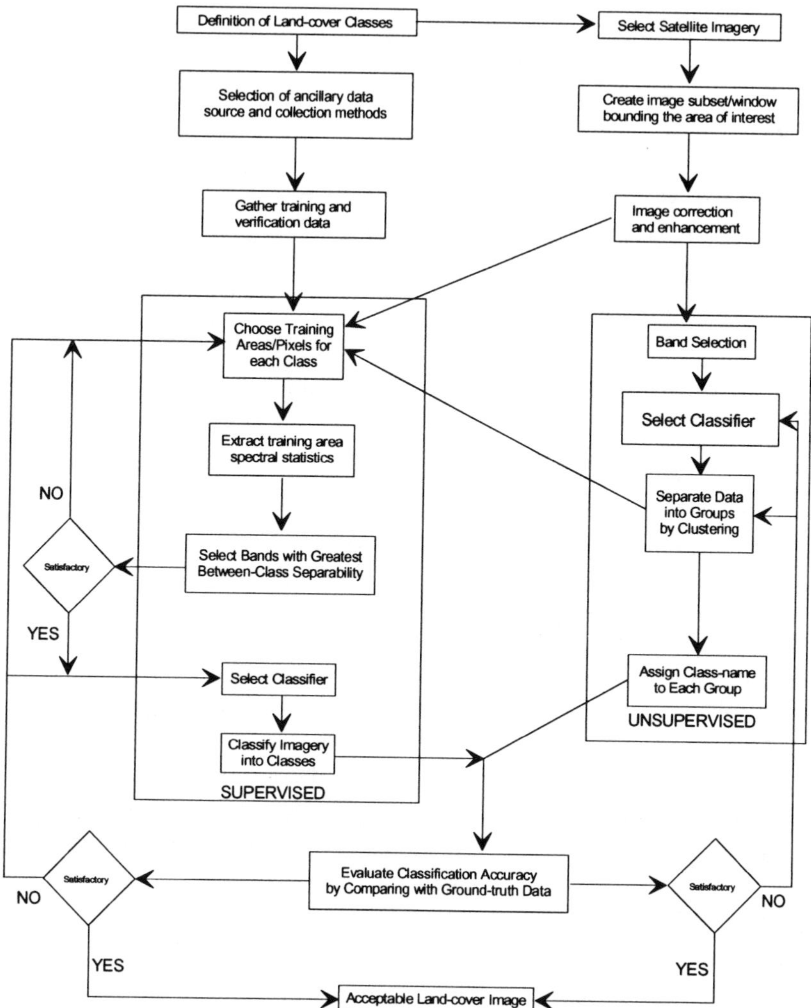

Figure 7.1 Basic steps for classifying images.

to real land-cover types. In this process many spectral classes can be assigned to a few land-cover types.

In *supervised classification* the researcher first locates areas within the imagery that, based on maps and field work, are known to belong to particular land-cover types. From the pixels within these known areas the spectral characteristics of each land-cover type is calculated. Each pixel

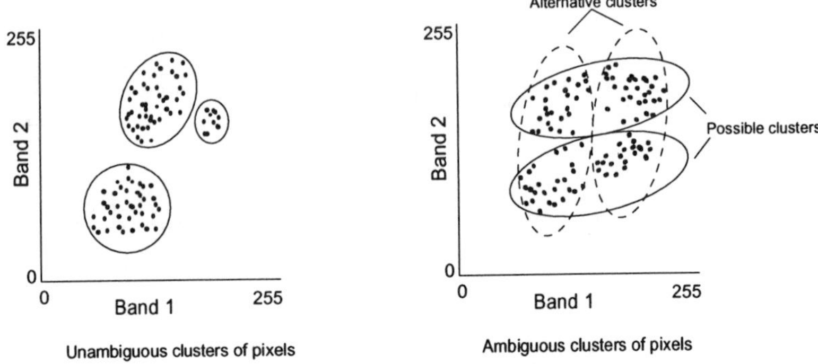

Figure 7.2 Pixel brightness values from two spectral bands plotted against one another. The results show separate (*left*) and ambiguous (*right*) clusters.

within the image is then assigned, by a computer, to the land-cover type that most closely matches its spectral characteristics.

Unsupervised Classification

Unsupervised classification uses statistical clustering techniques to combine pixels into groups (classes) according to the degree of similarity of their brightness values in each spectral band. For spectral cluster classes to represent a single landscape class exclusively, the multispectral distribution of brightness values for each landscape feature must be unique (discontinuous). In the two-dimensional example depicted in figure 7.2, we can see that if two spectral bands are plotted against one another, the pixels that represent unique landscape features might in some instances cluster together, and the distance between pixels that represent one feature would be less than that between pixels of two different features. Often, however, a two-band plot produces ambiguous information. The location of features in multispectral space is often not unique, and pixels could potentially be considered as members of several landscape features.

The spectral response (signature) of any feature changes from day to day with changes in sun angle, weather (rain, snow, etc.), atmospheric conditions (haze, smoke, etc.), tides, water levels, and human disturbance (farming practices, forest management, etc.). Vegetation phenology—the changes in leaf production, flowering, and fruiting that occur with all plants over time—is also an important factor in changing spectral response of a landscape. As a result, some features that should be distinguishable, at least theoretically, may appear to be identical within the imagery. For example, though conifer forest and wooded wetlands usually exhibit uniquely identifiable spectral responses, if the former lies in shadow, or the imagery was obtained soon after a heavy rain, these two landscape features may be indistinguishable.

Several algorithms can be used to generate spectral classes from multispectral imagery. Depending on the method and clustering parameters we choose, the number and composition of classes generated for a given image will vary. Identification of landscape classes is more like attempting to determine an individual's suit size based on height and weight, and less like clustering people into discrete categories such as sex, level of education, car ownership, and income. Figure 7.3 plots a hypothetical population of men according to height and weight. The swarm of data points is continuous, with no clear dividing lines. As clothes manufacturers cannot make custom tailored clothes for each man in the population, they must decide how many sizes to make and where they should draw the lines between sizes. Division of landscapes into feature classes is usually as subjective as the designation of who is small, medium, extra-large, tall or short. In most landscapes the interstices between patches of uniform landscape (oak or pine forest) are composed of a heterogeneous mix of species. We are left with the dilemma of how many classes to make, therefore, between pure oak and pure pine forest. Should we make one mixed forest class, or two, or thirty?

The number of clusters created from a remote sensing image, and the pixels that form the membership of each cluster, is very dependent on the clustering routine used and the mathematical parameters set within the program. The two most common clustering techniques are (1) iterative (or migrating means) and (2) agglomerative hierarchical clustering. *Iterative clustering* requires that the researcher select the number of clusters to be generated. The software program selects, arbitrarily, points within an *n*-dimensional space (one dimension for each spectral band being used to generate clusters) to represent the center point of each unique cluster. It then assigns pixels to the closest cluster in Euclidean terms. Once all pixels are assigned to the nearest cluster, the program calculates a new center point for each cluster, based on the mean location of all pixels within the cluster. The program then calculates the sum of squared error (SSE) for all pixel-to-center distances. The program now reassigns

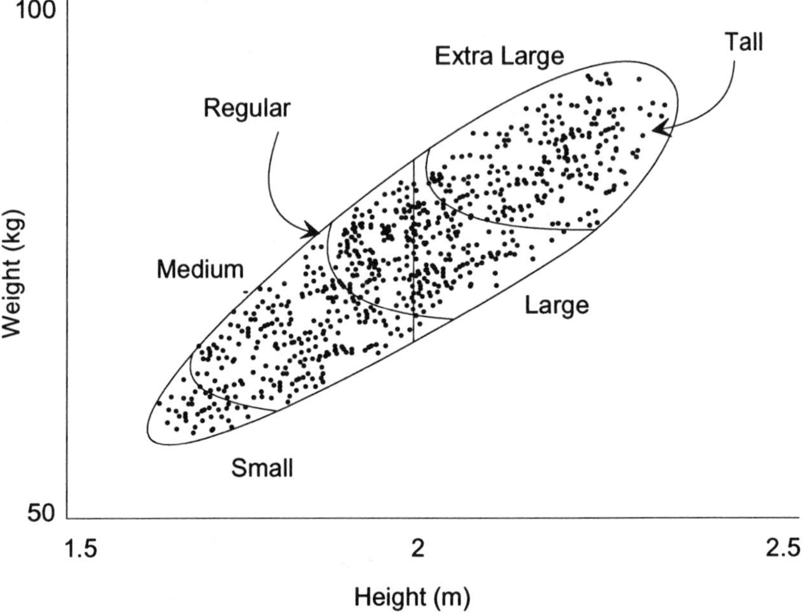

Figure 7.3 Difficulties in defining discrete landscape classes. Identification of landscape classes is more like attempting to determine an individual's suit size based on their height and weight, and less like clustering people into discrete categories such as sex, level of education, car ownership, and income. This figure shows a plot of a hypothetical population of men according to their height and weight. The swarm of data points is continuous with no clear dividing lines. As clothes manufacturers cannot make custom-tailored clothes for each man in the population, they must decide how many sizes to make, and where they should draw the lines between sizes. Division of landscapes into feature classes is usually as subjective as the designation of who is small, medium, extra-large, tall, or short. In most landscapes the interstices between patches of uniform landscape (oak or pine forest) are composed of a heterogeneous mix of species. We are left with the dilemma of how many classes to make, therefore, between pure oak and pure pine forest. Should we make one mixed forest class, or two, or thirty?

all pixels to the closest new cluster center. This iterative process—assigning pixels, calculating new cluster centers (migrating means), and reassigning pixels—continues until the total pixel-to-center distance is minimized (i.e., the sum of squared error is minimized). Figure 7.4 shows a two-dimensional example of the iterative, or migrating means, clustering method. With each iteration the cluster centers migrate according to the mean location of pixels within the cluster, and the pixels are reassigned to the closest new cluster center until the SSE is minimized. Notice that two pixels that were initially members of the upper right cluster switched membership to the lower left cluster by the last iteration.

Agglomerative hierarchical clustering is considered a better method than iterative clustering because it does not require a priori specification of the number of clusters to be created and because it can generate a fusion dendrogram that tracks the history of cluster mergings. The agglomerative hierarchical method of clustering starts by assuming that each and every pixel is a distinct cluster center. Subsequent steps systematically merge more and more neighboring pixels into larger clusters, according to a minimum distance-to-means criterion. (A distance-to-means criterion requires each pixel to be assigned to the closest cluster center; pixels that are closest to one another will form clusters.) The routine stops when all pixels are merged into one cluster. We can then determine objectively the most realistic set of clusters to assign pixels to, by looking for the longest distance between mergings, given the assumption that this stability represents robust (realistic) clusterings. As the clustering method starts with every pixel as a cluster center, the number of iterations (and the speed of the process) is a function of the size of the image. Small images with few pixels can be clustered much more quickly than can large images. Figure 7.5 presents an example of this clustering method.

Once the unsupervised classification has generated a set of spectral clusters (classes), we still have the task of labeling them so that they correspond to real landscapes on the ground. This is achieved by visually comparing the location of particular spectral classes to the location of landscape features shown on maps or other ancillary data for the area covered by the image. This process of assigning spectral features to a range of real landscapes (land-cover classes) often results in the merging of spectral classes. Table 7.1 shows how clustering of spectral data within an image tends not to generate spectral classes that have a direct one-to-one relation

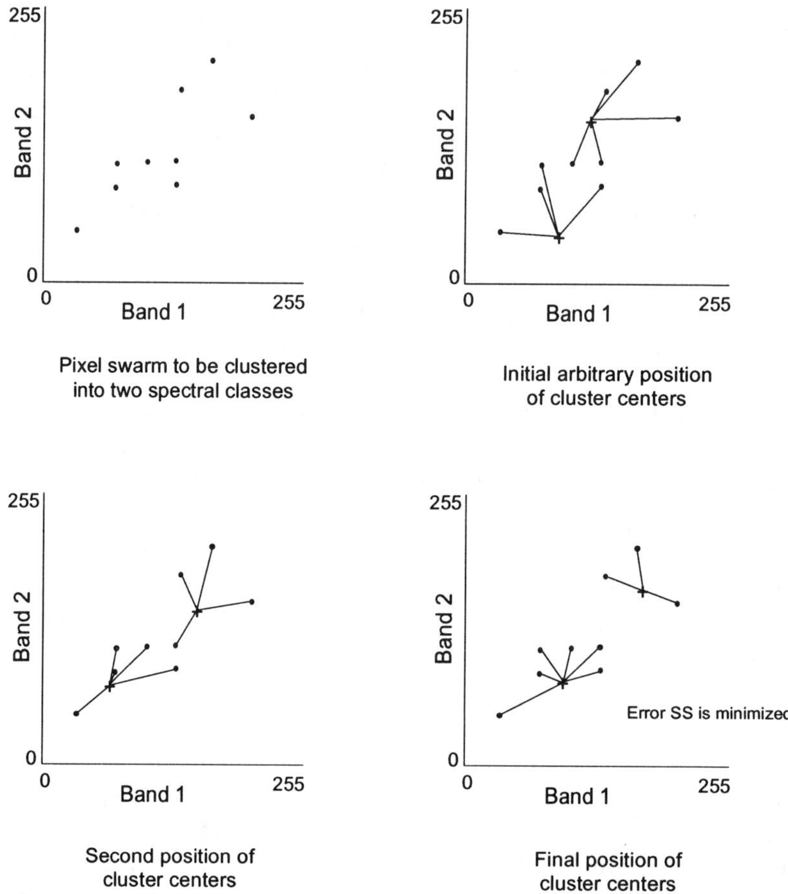

Figure 7.4 A 2-D example of the iterative (or migrating means) clustering method. With each iteration the cluster centers migrate according to the mean location of pixels within the cluster; the pixels are reassigned to the closest new cluster center until the total of all the distances of pixels to cluster center is minimized (i.e., the sum of squared error—SSE—is minimized). Notice that two pixels that were initially members of the upper right cluster switched membership to the lower left cluster by the last iteration. After Richards 1986:193.

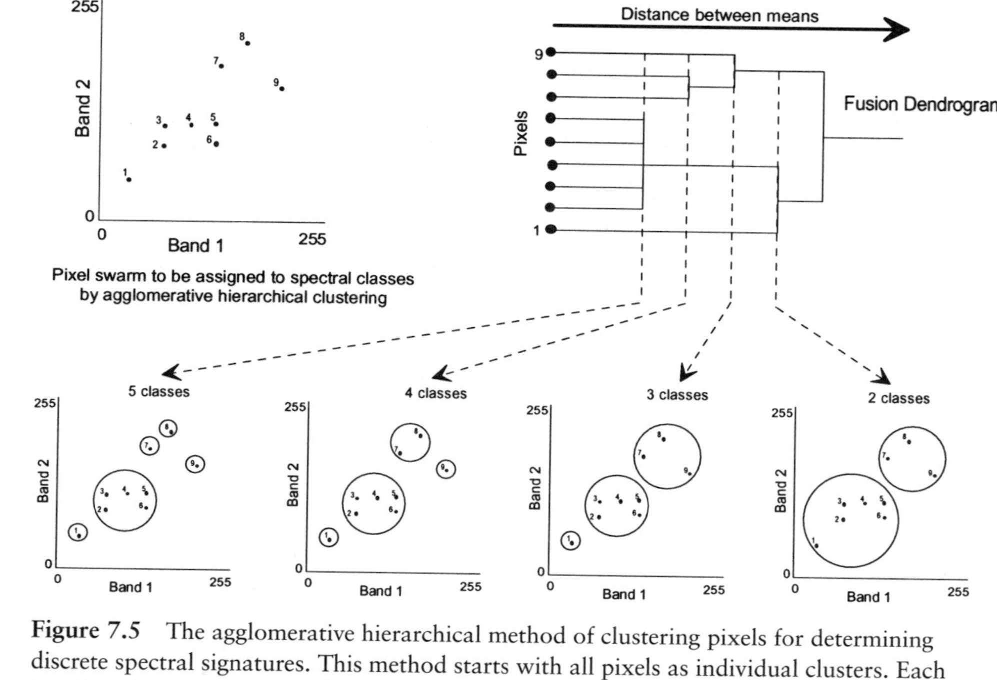

Figure 7.5 The agglomerative hierarchical method of clustering pixels for determining discrete spectral signatures. This method starts with all pixels as individual clusters. Each iteration clumps pixels together according to their proximity to one another. This method is preferred over migrating means as it does not require a priori determination of the number of clusters to be created, but becomes exceedingly slow as the number of pixels (i.e., the size of the image) to be clustered increases. After Richards 1986:201.

Table 7.1
Ambiguous Mapping of Spectral Classes Into Real Land-Cover Types

Migrating means clusters	Composition of spectral cluster	Researcher's land cover class
Six clusters		
1	Water	River
2	Forest	Old growth forest
3	Mixed forest and agriculture	Regrowth forest
4		Slash and burn farm
5	Agriculture	Village
6	Village and dirt road	Road
	Palm forest	Palm plantation
		Palm forest
Eight clusters		
1	Deep water	River
2	Water and river bottom	
3	Forest	Old growth forest
4	Mixed forest	
5	Regrowth forest	Regrowth forest
6	Agriculture	Slash and burn farm
7	Village and dirt road	Village
8	Palm forest	Road
		Palm plantation
		Palm forest

to land-cover classes on the ground. Pixels within some spectral classes could be members of more than one land-cover class. Similarly, more than one spectral class may need to be merged into a single land-cover class. It is wise, therefore, to err on the side of generating too many rather than too few spectral classes initially. If, however, a spectral class is found to represent more than one real landscape feature, that class can be extracted from the imagery; by using the same clustering method, it can then be split into several groups that may be unambiguously assigned to landscape features of interest.

Unsupervised classification is undertaken most often as a guide to collecting ground data before a supervised classification is undertaken. For example, after completing an unsupervised classification of an image to separate areas that have unique spectral characteristics, a researcher would visit, in the field, a few examples of each land-cover class important to the study question and would determine how these sites relate to the unique spectral classes generated by the unsupervised classification. Lack of a one-to-one relationship between land-cover and spectral classes may encourage the researcher to modify the number and characteristics of land-cover classes, or to reevaluate the type of remote sensing imagery to be used in the study. Unsupervised classification may also prove useful as a last resort, if ancillary data are insufficient for undertaking a supervised classification.

Supervised Classification

Unsupervised classification generates a unique set of spectral classes that must subsequently be assigned to represent particular landscape features. Supervised classification, however, converts the spectral data contained within remote sensing directly into thematic land-cover information. To perform a supervised classification, the user identifies homogeneous regions within the image that represent unique known landscapes. These areas (*training sets*) are used statistically to generate spectral signatures (responses) characteristic of each landscape type. A digital classifier then compares the spectral signature of each pixel in the image to the training set signatures, thereby determining to which landscape type each pixel is most likely to belong. The image generated is, consequently, a thematic land-cover map of the area.

For supervised classification of remote sensing imagery to be effective, we must be able to do the following:

- Select or develop a thematic land-cover classification with mutually exclusive and exhaustive classes that can represent all landscape types likely to exist within the image.
- Select pixels (training sets) within the image that represent homogeneous patches of each land-cover class listed in the classification and present in the area covered by the image. Training sets can be selected using maps, aerial photographs and videography, field visits, and photo interpretation of enhanced single-band and color-composite imagery.
- Detect and delete unrepresentative (outlier) pixels within training sets.
- Extract the characteristic spectral signature from a random sample of the pixels contained within the selected training sets for each land-cover class.
- Determine, by examining the spectral signature statistics and displaying bivariate plots of brightness values, the spectral bands that in combination manifest the greatest between-signature separation of all land-cover types.
- Select the most appropriate classifier for the quality of training set data available and the size of the image to be classified.

- Apply the classifier to assign each pixel within the image to a land-cover class, using the training set spectral signature statistics as class archetypes.
- Establish quantitatively the accuracy of the classification.

Selecting a Land-Cover Classification Scheme

Depending on what region of the globe you are working with, you may or may not be able to use a tried and tested land-cover classification scheme. Examples of such schemes include the ones devised by Cowardin et al. (1979) for terrestrial and coastal wetlands, by Malingreau (1977) for landscapes characteristic of urban and rural Indonesia, and by Andersen et al. (1976) for North America. The Cowardin system has been used by the Fish and Wildlife Service to classify and map all wetlands throughout the United States that have been visually detected in high-resolution IR aerial photographs. The Andersen and Malingreau systems were specifically developed to classify spectral features visible within satellite remote sensing imagery.

If a suitable classification scheme is not available, then one must be developed—preferably by modifying an already proven classification. Adopting or modifying existing classifications is preferable to devising completely new schemes for several reasons. Developing effective classifications is time consuming. A new, esoteric classification may be used only once; it precludes comparisons with other studies and inhibits the sharing of data.

Bailey et al. (1978) provide a review of classification schemes. They conclude that a useful classification must be flexible, and that flexibility is best achieved by a hierarchical structure, with each level composed of mutually exclusive and exhaustive land use or land-cover categories (classes). Though creation of an exhaustive classification scheme may require time-consuming characterization of features irrelevant to the particular study question, that work is essential for avoiding the possibility of erroneously labeling uncharacterized areas (pixels) during the digital classification stage of the analysis.

The goal of developing a useful classification is to accurately label spectral classes distinguished within the imagery according to their corresponding land uses, land covers, or ecosystem functions observable on the ground. *Cover types* within a land-cover classification can denote individual plant species (e.g., white pine plantation) or unique land forms (e.g., open water). Cover types can also entail mixtures of species (e.g., hardwood forest, scrub-savanna, montane forest) or mixtures of land forms (e.g., urban, agricultural, airport). It is important, therefore, to remember that scale has an enormous impact on the perceived homogeneity or heterogeneity of landscapes, and thus the purity of pixels, spectral features, and land-cover types. Because most landscapes are mixtures of smaller features, the skill is to name mixed feature classes that are meaningful both within the imagery and in reality.

Preparing Training Sets

Once a land-cover classification system has been chosen or devised, selection of training sets for each class within the classification can begin. The goal of this phase is to locate within the imagery the pixel blocks that correspond to land-cover types of interest to the user. This phase is probably the most important part of supervised classification, for it is the spectral signatures extracted from these sets that will determine the overall classification accuracy, and thus the utility of the final thematic map. Care, therefore, should be taken in selecting training sets that represent typical (normative) examples of each land-cover class, and avoiding uncharacteristic, mixed, or atypical areas.

As all landscapes change over time it is, of course, very important to use training set information that was recorded contemporaneously with the imagery to be classified. That is, the imagery and training set information should be gathered as close to the same time as possible. This is particularly important for rapidly changing landscapes—what was forest during training set data collection may have been cleared for cattle pasture by the time the imagery is obtained if the hiatus between the two events is long. Similarly, ancillary differences in soil type, rainfall distribution, and topography may result in the same class of land cover showing a different spectral signature in different locations within the area covered by the imagery. It is therefore important not only to understand the factors that may produce geographic variation in class signatures but to take steps to mitigate these effects. Mitigation can be achieved by the stratified random selection of training sets through the area. (This method will be discussed in the next section.)

Another way to improve the usefulness of ancillary information is through a preliminary unsupervised classification of the imagery. This step can be exceedingly useful in identifying the range of unique spectral features contained within the scene. The unsupervised classification can then be compared to land-cover maps to identify whether areas known to be of the same cover type are represented in the image as the same or as different spectral classes.

In areas with numerous visual clues—such as road intersections, islands, and river confluences—it may be relatively easy to match landscape features on the ground with features within the imagery. In these instances it may not be necessary to geometrically correct the imagery in order to locate training sets accurately. More often, however, selection of training sets is easier if the imagery is transformed, using ground control points and nearest-neighbor sampling, to a standard map projection coordinate system such as UTM. Once the imagery is geometrically corrected, we can very easily overlay landscapes of known coordinates within the image.

The two most common sources of land-cover information are published maps and field surveys. Field surveys are considerably more expensive, but they often provide greater descriptive detail. Extraction of training sets from maps is most easily achieved by digitizing the boundaries of unique land-cover types and then superimposing these boundaries onto the imagery. A digital map that contains only the boundaries of features is called a *vector image* (figure 7.6).

Most image-processing software include digitizing routines that allow boundaries of land-cover features on maps or aerial photographs to be traced manually, using a cross-hair cursor attached to a digitizing tablet or table. The geographic location of the digitized points, lines, and polygons (boundaries of areas) that represent features within the map or photograph can be digitized using the same coordinate system as the geometrically corrected raster image. In this way, the vector and raster images can then be superimposed and eventually compared.

Spatial errors within digitized data are a function of the scale of the map. Following a 1mm boundary line on a 1:10,000,000 map, for example, will result in larger displacement errors than following the same 1mm line at 1:10,000 scale. Spatial errors introduced during digitization can also reflect inaccuracies in the original map and lapses in the technical skill of the human digitizer.

Maps can also be digitized automatically by using a scanning densio-

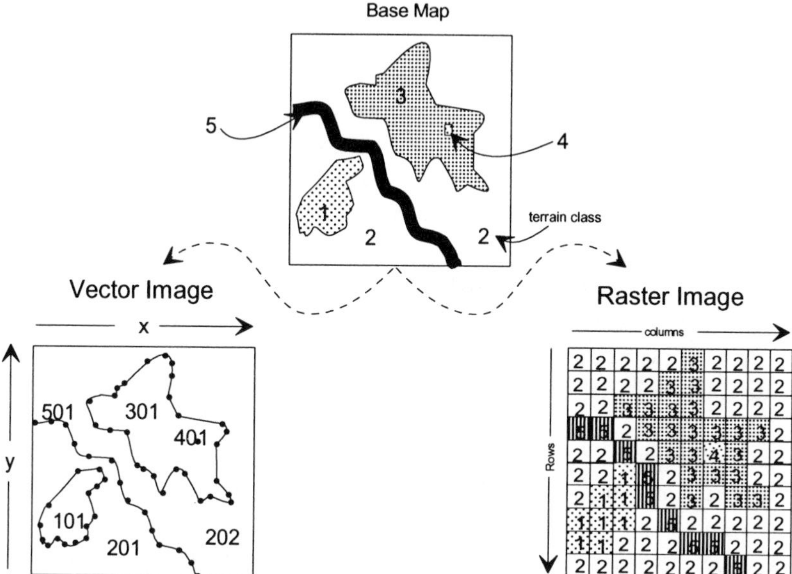

Figure 7.6 Base map and the vector and raster images derived from it. In a vector image, each base map feature is represented as an individually labeled point, line, or boundary of a polygon. In this example, point 4 in the base map becomes a vector image point with a unique label (401), a terrain class identifier (4) and a pair of *x,y* coordinates designating its location within the image. Line 5 is represented by the label 501, class identifier 5, and fourteen *x,y* coordinates. The area 3 becomes polygon 301 of terrain class 3 in the vector image, the boundary of which is defined by twenty-seven *x,y* coordinates. In the raster image, the base map is represented as a matrix of regular cells; each cell has a row and column coordinate and a label that corresponds to the underlying terrain class. One major difference between vector and raster data is in feature labeling. In the raster image the cell labeled 2 in the lower left corner of the matrix is indistinguishable from the cell labeled 2 in the lower right corner of the image. In contrast, an area within object 201 in the lower left of the vector image, though of the same class, is distinguishable from all areas that lie within object 202. An area measurement for terrain class 2 would generate two values in the vector image (one for each discrete polygon—201 and 202), but only one value in the raster image, as all cells labeled two are equivalent.

meter (scanner), which converts the map into a *raster* (bitmapped) image composed of pixels (figure 7.6). Computer ability to translate a base map into a vector map, however, is still in its infancy; translation errors have to be corrected laboriously by hand. This difficulty is not surprising, as most lines on a map intersect with other lines and labels. A line-following program will thus have difficulty determining which line to follow at an intersection; a vector that starts off following a road may end up following a contour, utility line, railway, or city name. Very often it is faster to digitize map information by hand, rather than to attempt automated vectorization of scanned maps followed by laborious correction. The accuracy of converting a scanned map to a vector image can be improved greatly by following these steps:

1. Photocopy the original map using a redtone copier (available at most office copy shops), reducing or enlarging the map to best fit the scanner.
2. Using a black marker, trace over the lines to eventually be converted to vector format.
3. Scan the redtone photocopy, if possible as a 1 bit (binary) grayscale image, and save the digital image in TIFF or PICT format. By adjusting the scanner, only the black marker lines will be resolved and the red background will be left blank.
4. Import the TIFF or PICT raster image into the image processing software.
5. Geometrically correct the imported raster image.
6. Mosaic map segments together if the complete map could not be scanned in one piece.
7. Vectorize the complete raster image; most image processing software contain routines to do this.

As software gets smarter, and as it comes to include optical character recognition and artificial intelligence methods, automatic scanning will gradually become a reality. But often we need not wait. Topographic and street maps of many locations around the world are now available in digital vector format. (*Software for Science,* produced by SciTech in Chicago, is a useful source of digital map data.)

One need not have access to a digitizer and scanner in order to extract clusters of pixels to use as training sets. Most image-processing software allows for on-screen digitizing of chosen areas—and these can become the training sets. On-screen digitizing works well when land-cover patches are large, and when sufficient visual clues are evident in the image and on the map to permit the unambiguous location of training sets (figure 7.7).

Once training set polygons are created for each cover type, they can be used as a mask to extract pixels assumed to be representative of each cover type. These cover type–specific pixel sets are then available for statistical analysis to determine the characteristic spectral response (signature) of each cover type.

Setting Up a Field Survey

Before the advent of aerial and satellite remote sensing, field surveys were in fact the only source of information for describing and mapping landscapes. Field surveys are, however, exceedingly time consuming and costly, so they must be undertaken efficiently. Until remote sensing imagery was readily available, an understanding of the geographic distribution of a particular landscape was based on just a few localized studies. As a result, any such understanding remained rudimentary. Because patches of the same landscape features are assumed to share a similar spectral response regardless of their geographic location, a major benefit of remote sensing image analysis is the capacity to regionalize land-cover distribution accurately from ground information obtained from only a few localized field surveys.

An understanding of the spectral response of vegetation, water, and soils can help to characterize, in very coarse terms, landscape features within remote sensing imagery, even in the absence of field surveys or other ancillary information. However, detailed land-cover maps, and an assessment of the accuracy of these maps, is impossible without appropriate reference data that include field surveys. Reference data (maps, field surveys, aerial overflights, etc.) are essential for training human interpreters and digital classifiers how to relate landscape features on the ground with spectral features within the imagery. Reference data are also important for testing the accuracy of visual or digital classifications.

For many remote sensing studies, valuable land-cover and land use information to guide selection of training sets is available from accurate, up-to-date topographic, soils, and vegetation maps. Where this is not the

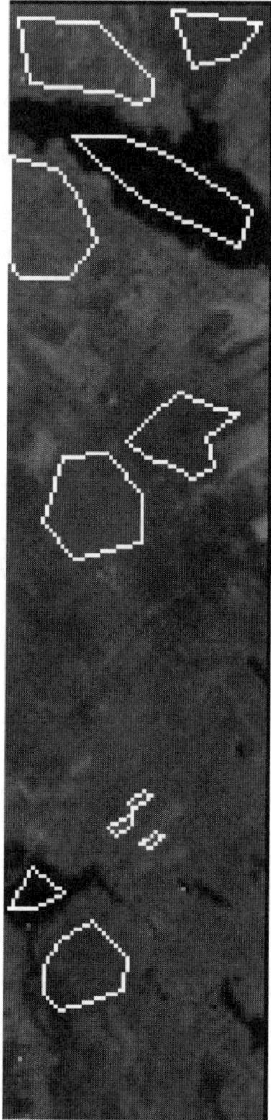

Figure 7.7 Training set polygons in vector format overlying a raster image. The vector format polygons delimiting the boundaries of the training sets within the underlying raster image were digitized using an on-screen cursor and a mouse. The location, size, and shape of training sets are determined by the researcher's personal knowledge of the area.

case, field surveys are essential and not merely desirable. Field surveys are expensive, in terms of both time and money. A cost-effective approach for obtaining the field data needed for completing an accurate supervised classification is thus imperative. Our discussion in this section focuses primarily on designing an appropriate protocol for collecting ground data for a satellite land-cover classification. Nevertheless, much of the information on sampling strategies presented here is relevant to other objectives and to using published maps as sources of reference data. Several examples of using field sampling methods in remote sensing studies are presented in the next chapter.

The goals of field surveys are to

- visit areas (sites) that represent examples typical of each land-cover class present within the area;
- describe the composition of the area in terms relevant to the spatial and spectral characteristics of the imaging sensors; and
- record the geographic location of the area with a level of precision appropriate to the pixel size of the imagery and the surface area of the field site.

If the field study is successful, we will be able to locate areas within the imagery that represent known (described) land-cover classes. The spectral characteristics associated with these areas can then be used as archetypes by which to classify the whole image. Ancillary data such as ground surveys are essential to generating classified images and to verifying the accuracy of land-cover maps derived from remote sensing data.

In setting up a field survey, one needs to answer several key questions:

- What is the optimal timing for field surveys, relative to the time that remote sensing imagery was or will be acquired?
- How does one decide upon a typical example of a land-cover class?
- How many sites need to be visited to constitute a statistically representative sample?
- How large should sample sites be?
- What types of information should be collected at each site?

Choosing a Time for Data Collection

Assuming that the area of interest is accessible (i.e., that it is not an isolated tropical forest during the heavy rains, or Antarctica during the winter), field data can be collected at any time. Data collection at or very close to the time the remote sensing imagery are recorded is desirable, particularly if sun angle, phenological, weather, and land use changes are highly dynamic.

Synchronous or near synchronous data collection requires that field data collectors are standing ready for a time when weather conditions permit successful acquisition of the remote sensing data. In most situations absolute synchronicity is not possible, unless data are being gathered over a perennially cloudless area or near to the survey team's home base. In addition, field data collection before, or synchronous with, image acquisition requires that archival reference information such as maps or aerial photographs are available to guide selection and location of field sites. This is certainly not the case in many isolated regions of the tropics, where biological diversity is greatest.

If information about the area is very sparse or out of date, it is difficult, a priori, to identify and locate field sites to visit that are typical of each land-cover type present within the area—or at least those likely to be of interest to the researcher. Under these conditions, it is essential to first obtain the imagery, and then, as soon as possible thereafter, collect the field data. Field surveys conducted after image acquisition must, of course, attempt to assess what changes in the landscape have occurred since the imagery was obtained (e.g., leaf cover of deciduous vegetation, land clearing for agriculture, soil moisture, flooding).

Preliminary Image Analysis to Guide the Field Survey

In general, it is good practice to undertake a preliminary image analysis before going out into the field. By viewing the remote sensing data as single-band and color-composite images, the researcher begins to get a much better sense of the spatial and spectral detail available within the imagery. This is exceedingly important because it instills realistic expectations as to what features the imagery can and cannot resolve (identify) readily. Having a clear sense of the spatial and spectral resolution of the imagery within a particular geographic context is very important when making decisions as to the number of categories to include in the land-cover classification scheme, and the appropriate size for field survey sites.

The most effective means for identifying the size, shape, and distribution of features that represent the spectral variation contained within the image is to conduct an unsupervised classification of the data. The spectral class map generated can then be the starting point for designing an appropriate sampling scheme that stratifies field surveys among all the spectral features of interest. By visiting and describing an appropriate sample of each of the spectral classes distinguished in the unsupervised classification image, we can be confident that we have characterized the major land-cover types identifiable, at least spectrally, within the imagery. It is important to remember that spectral classes generated using an unsupervised classification of the imagery may not correspond unambiguously with the land-cover and land use types that can be detected on the ground. In some cases one or more spectral classes may represent a single land-cover type; conversely, a single spectral class may include several land-cover types that the researcher would like to distinguish (table 7.1). The former can be resolved by clumping spectral classes. The latter may be resolved by attempting to split the ambiguous spectral class into several classes, by using fine-grained unsupervised cluster analysis methods or by using additional data layers such as terrain elevation, slope, and aspect. There is, of course, no assurance that the land-cover or land use types of interest are in fact spectrally or spatially separable.

Preliminary examination and analysis of imagery before field work is, we believe, the most effective way of ensuring that field data collection is cost effective.

Determining the Composition, Size, and Number of Field Sites

Pixels within scanner-based remote sensing imagery represent the total reflected and emitted radiation from all terrain features within an area equivalent to the IFOV (instantaneous field of view) of the sensor. Field data collection aims, therefore, to characterize the landscape within a given area, rather than at a specific point in the landscape. Similarly, as one of the initial goals of most remote sensing analyses is to create a map composed of mutually exclusive land-cover or land use types, field data collection must target areas that represent each cover type found both within the classification scheme and within the imagery. To accommodate these two objectives, field sites must be

- internally homogeneous, such that the vegetation, soils, and land use features of which they are composed are representative of only one land-cover category;
- large enough to be unambiguously located within the imagery, to have pure (unmixed) pixels, and to be found and described on the ground; and
- sufficiently numerous so that they account for the spectral response range of each land-cover category.

Though industrial-scale, single-crop forms of agriculture provide good examples of internally homogeneous landscapes, comparable examples are uncommon in nature (except, inter alia, monodominant forests, non-vegetated hot and cold deserts, and *Spartina* salt marsh). Landscapes rarely change abruptly. Rather, they blend into one another, making it difficult to determine where one community feature ends and another starts. Most landscapes are likely to include mixtures of species and to have variable soil exposure and topography.

How much variability within a site can a researcher tolerate before the site is no longer considered internally homogeneous? As changes in surface cover, aspect, and slope are the major reasons for variability in the spectral response of a landscape, Justice (1978) proposes that an internally homogeneous site is one that (a) possesses a single cover type over more than 85% of the site, or has a uniform spatial distribution of cover types throughout the mixed site, (b) has no more than a 22.5% variation in aspect, and (c) has no more than a 20% variation in slope over no more than 20% of the site area.

Finding field sites that meet these internal homogeneity criteria is a goal worth striving for. All field surveys, as a matter of course, should measure the cover, aspect, and slope characteristics of prospective sites. However, depending on the degree of heterogeneity of the landscape and the spatial resolution of the sensor, it may be exceedingly difficult to find sites that are truly internally homogeneous. For example, if the spatial heterogeneity of the landscape is greater than the size of a pixel, no sample site is likely to be internally homogeneous—even if we could visit and describe the location of a single pixel within the imagery.

In any situation where the brightness value of the majority of pixels is composed not of a single cover type but of two or more cover types we have a "mixed pixel" problem (figure 7.8). Mixed pixels occur in images

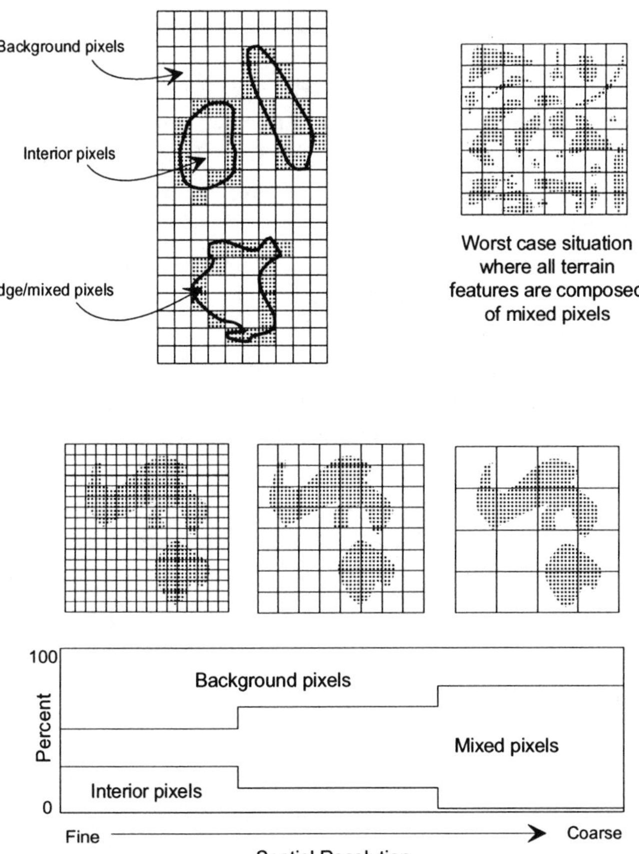

Figure 7.8 Mixed pixels caused by landscape patchiness or coarse spatial resolution of the sensor. Note in the upper left how the shape of a feature influences the proportion of edge or mixed pixels.

of very spatially heterogeneous landscapes, at the edges of homogeneous features, and along long linear features such as roads, rivers, and railways. For example, small open-water wetlands scattered across the landscape may, depending on the resolution of the sensor, appear only as mixed pixels, and may thus remain undetected.

Mixed pixels occur in all classification schemes using raster data, so the objective is to minimize their abundance. Ways to limit the mixed-pixel problem include (1) obtaining higher spatial resolution imagery, (2) creating land-cover classes to represent the most common mixed pixels, and (3) selecting larger sample sites for developing training sets in uniform landscapes, so that more pixels are pure pixels. A good rule of thumb is that field sites should consist of 3–5 pure pixels (0.2 hectares for SPOT HRV, and 3 hectares for Landsat MSS), which, depending on the shape of the site, could translate into an overall area as large as 20–40 pixels.

Determining the total number of sites to visit broaches the dilemma of all research: the desire for a sufficiently large sample size, but the constraints of limited time and money. A rough guide for navigating this dilemma is that enough sites should be visited and described such that $50x$ pure pixels are available to estimate the spectral response for each land-cover class, where x is the number of spectral bands to be used to estimate class spectral signatures. In highly heterogeneous landscapes, as many sites as possible is a useful maxim, because statistical analyses tend to be robust, even when assumptions of independence are violated, if they are based on a large number of samples (i.e., a large number of error degrees of freedom).

Selecting an Appropriate Sample of Field Sites

A major advantage of using remote sensing data is the ability to extrapolate information on the spectral characteristics of local terrain features over the whole region covered by an image. The accuracy of a land-cover map generated from remote sensing images will depend, therefore, on the validity of the sample (ground training and verification areas) from which the regional extrapolations are made. Selection of sample points and areas can be accomplished by either subjective (purposive) or objective (probability, random) sampling.

Subjective sampling determines training sites based on the knowledge of the researcher or local informants, with the belief that the sites are representative of the cover types present within the landscape. Subjective sampling may serve to reduce the time and expense of conducting ground surveys, as the most accessible sites are likely to be chosen by an individual who knows the region well. Yet, one can never be sure whether subjective sampling is truly representative of the cover types within the landscape. All statistical methods used to classify digital remote sensing imagery assume that training sites (samples) are selected randomly and are not correlated with one another. As subjective sampling violates at least the first assumption, the validity of the classification will be questionable, and the accuracy of the regional, extrapolated survey cannot be established.

Objective sampling uses a set of standardized procedures to select sites from within the region ("population," in statistics terms) such that each site has a known probability of being selected. The simplest method is random sampling, whereby all points or areas within the region have an equal probability of being included in the sample. For random sampling, the region is overlaid with a matrix of equal-area cells. Sample cells are then selected, using pairs of random numbers to represent the cell's row and column position.

To ensure that all regions of the image are sampled evenly, but to still ensure that all areas have an equal probability of being sampled, an *unaligned systematic random sampling* scheme can be used (figure 7.9). If ancillary information (maps, remote sensing imagery) about the distribution of landscape features over the area covered by the imagery is not available, it may be necessary to resort to a simple random or systematic sampling protocol. When ancillary information is available in the form of an unsupervised classification of remote sensing imagery or maps, it is possible to stratify the location of field sites such that each spectral class is sampled in proportion to its relative area. A *stratified random sample* is helpful in that it ensures that all land-cover types are visited regardless of how rare they are, and larger areas that may be more internally heterogeneous are sampled at a higher rate.

To reduce the travel time between randomly selected sites, a *random nested sample* can be used. In this multistage process the region to be sampled is subdivided into a number of equal areas, and a sample of these selected. These subareas are themselves subdivided, and a further sample is selected from them. The selected area subdivision and resampling continues until field sites of suitable size are selected. Nested sampling can also be the basis for a hierarchical approach that takes advantage of ancillary information of different scales. For example, AVHRR 1km imagery could be used to determine suitable woodland and grassland regions; within these regions Landsat TM imagery could be used to locate areas of deciduous and coniferous forest, and within these areas IR aerial photography could be used to select the final deciduous wooded wetland and upland field sites.

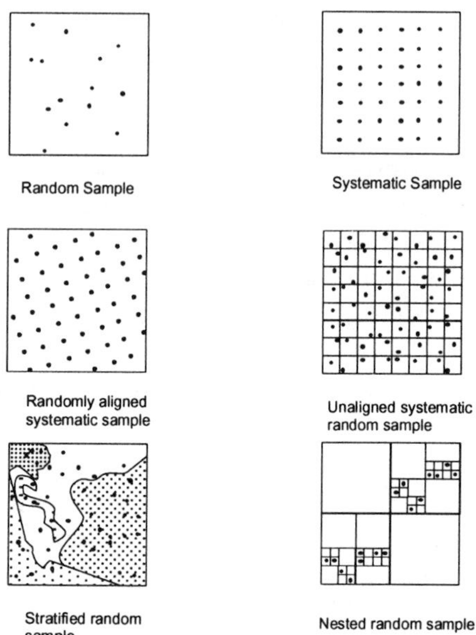

Figure 7.9 Random sampling options for collecting ground truth data. After Townshend 1981:42.

Executing a Field Survey

Field data collection as a vital aid to remote sensing image analysis entails two major objectives. The first is to obtain information on the reflected and emitted radiation characteristics of the site; the second is to describe the site's biophysical characteristics at a macro rather than micro level. Once we have selected our field sites, what should we measure once we get there and how should we measure it?

What to Measure and How to Measure It

Spectrophotometers provide continuous spectral readings across a given spectral range. Band-pass radiometers provide spectral readings within discrete spectral bands that often match those of satellite multispectral sensors. Because both are used on ground-based platforms in the field, they are useful for measuring the spectral radiance of features without having to worry about atmospheric interferences.

Using a hand-held or boom-mounted spectrophotometer (sometimes also called spectroradiometer or radiometer) to gather reference data on the changing spectral response of landscape features under different atmospheric, weather, and phenological conditions is essential to enhancing understanding of the information content of remote sensing imagery. Such reference data are also important for extracting the direct biophysical properties and quantities from the imagery and for designing future sensors (Sellers et al. 1990). Though field spectrophotometry data do provide the user with hands-on information about the spectral response of the landscape features under a range of conditions, most multispectral instruments are expensive. For example, an SE590 spectroradiometer had a 1995 list price of about $15,000, sold by Spectron Engineering in Denver (303-733-1060).

The very small field-of-view of hand-held or boom-mounted terrestrial radiometers requires that multiple measures be taken to estimate an average spectral radiance for the site. Field radiometry is, however, generally impractical in wooded areas, as elevating the radiometer above the canopy can be both time consuming and expensive. Field spectrophotometry does provide detailed information on the spectral radiance characteristics of homogeneous and heterogeneous landscape features, and it is thus very useful for deciphering the information within remote sensing imagery. In contrast, laboratory spectrophotometry on leaves of individual species is most useful in theoretical rather than applied research.

Another category of field measurements targets biophysical characteristics, such as slope, aspect, soils, and vegetation of the site. These attributes are most influential in determining the spectral radiance of the site. Slope and aspect are important because, when considered in conjunction with the angle and azimuth of the sun or radar antenna, they determine whether a particular site will be illuminated directly by the radiation source or be in partial or complete shadow.

The proportion of exposed soil and rock in the site should be estimated. Soil structure, texture, organic matter, color (using the color charts published in Munsell 1975), and moisture should all be determined qualitatively, if not quantitatively. Vegetation cover and composition should be characterized only to the extent that it elucidates the likely spectral response of the field site. Data collection should focus, therefore, on the relative canopy closure or leaf cover of trees, shrubs, saplings, and herbs and grasses. These data can be measured directly, or they may be

Table 7.2
Field Site Description Template for Paper or Digital Form

Collector:____________________ Site#:__________ Image #:________ Date: ______ Time: _____

GPS Location LAT N/S____.____.____ LONG E/W____.____.____ Altitude:_____ PDOP:_____

Site Photographs: Roll___ Frame____ Roll____ Frame____ Roll____ Frame_____Roll____ Frame____

Photo Descriptions:____

Cover Type (Select 1 type) ________
- 1 Forest
  - 1.1 Closed forest (tree cover >50%)
  - 1.2 Open forest (tree cover 10-50%)
- 2 Shrubland (tree cover <10%, shrub cover >50%)
- 3 Open vegetated
  - 3.1 Savanna (trees <10%, shrub <50%, grass >40%)
  - 3.2 Herbland (trees <10%, shrub <50%, herbs >40%)
  - 3.3 Agro-pastoral land
- 4 Nonvegetated
  - 4.1 Open water
  - 4.2 Bare rock
  - 4.3 bare soil
  - 4.4 Buildings
  - 4.5 Other__________________

Hydrotopography (Select 1 type) ________
- 1 Hills (average >15% slopes)
- 2 Upland (flat to rolling, <15% slopes)
- 3 Flooded (standing water <50% of the year)
- 4 Swamp (standing water >50% of the year)

Dominant/Mixed Type ______________________
Dominant species (>75% cover) OR note MIXED

History (Select 1 type) ______________
- 1 Mature forest
- 2 Recent secondary forest
- 3 Abandoned field, farm bush
- 4 Kept clear or cropped
- 5 Other______________________

Slope to nearest 10° ____________

Aspect to nearest 10° ____________

Usage (Select several if needed) __________
- 1 None evident
- 2 Tree felling
  - 2.1 Slight (<10% cover affected)
  - 2.2 Moderate (10-50% affected)
  - 2.3 Severe (>50% affected)
- 3 Grazing
- 4 Subsistence and local market agriculture
- 5 Commercial agriculture
- 6 Other____________________________
- 7 Unknown

Cover/Closure ______________
score 0-5 for each type
0 1=>0-10% 2=>10-25% 3=>25-50% 4=>50-75% 5=>75
estimate cover for:
Trees
Shrubs & saplings
Grasses & herbs

Soil
Texture ________________
Color ___________________
Moisture _________________

inferred from species composition and abundance. Table 7.2 provides an example of the type of data to gather for a field site.

Not only is it important to determine the field site's biophysical characteristics, it is valuable to assign each field site to one of the categories of the land-cover classification selected or devised for the remote sensing study. Terrestrial and low-altitude photography and videography are exceedingly useful for comparing and contrasting field sites, and subsequently for verifying field site categorizations. Unless the field site is exceedingly small, it will be necessary to record slope, aspect, soil, and vegetation characteristics at several locations within the site to ensure that the area is described accurately. Point samples within the field site can be determined most efficiently by selecting one or several randomly oriented transects, and sampling at even or random intervals along them.

Recording the Location of Each Field Site

Several methods can be used to determine geographic location when visiting a field site. The most basic is using a map and compass to triangulate the site's position relative to visible landmarks (the marine equivalent is using an accurate chronometer and astrolabe or sextant to determine latitude and longitude). This method may provide precision sufficient for coarse resolution imagery, such as AVHRR, under optimal conditions (i.e., visible and unambiguously identifiable landmarks), but it is unlikely to be useful for fine-scale studies. Recently an easier and much more accurate method of determining geographic location has become both available and affordable.

The global positioning system The global positioning system (GPS) has been developed by the U.S. Department of Defense to replace the older Omega, Loran-C, and Transit navigation systems. Those earlier systems were inaccurate, of only limited geographic coverage, or had long lag periods between position fixes. Now that all twenty-four NAVSTAR (Navigation System with Time And Ranging) satellites are in orbit, the GPS system provides accurate navigation and geographic location twenty-four hours a day anywhere on the globe.

The NAVSTAR satellites circle the earth in orbits at an altitude of 20,200km, with a twelve-hour period. NAVSTAR has an orbital geometry of six planes, each inclined at 55° and with three satellites in each plane (six satellites are used as "spares"). This configuration enables reception of direct line-of-sight navigation signals from at least four satellites at any point at or near the earth's surface at all times. Each satellite transmits a coarse acquisition (C/A) navigation signal that provides users with location accuracy ranging from 15m (using one GPS receiver) to a centimeter or less when two GPS receivers are "connected" in differential mode. Simultaneous monitoring of three satellites gives two-dimensional (latitude and longitude) position when altitude is known. Four satellites (figure 7.10) provide complete three-dimensional positioning.

The GPS determines the location of a field worker on the ground or in a plane by measuring the distance from the user's GPS receiver to a group of NAVSTAR satellites, whose orbital position is known exactly at all times of the day and night. Distance from a satellite is determined by

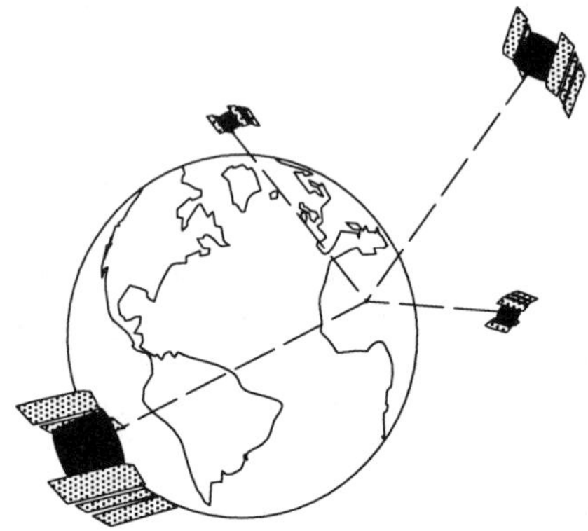

Figure 7.10 Accurately determining locations of field data collection sites using the global positioning system. Each NAVSTAR satellite transmits a coarse navigation signal that provides users with location accuracy ranging from 15m (using one GPS receiver) to a centimeter or less when two GPS receivers are "connected" in differential mode. Simultaneous monitoring of three satellites gives two-dimensional (latitude and longitude) position when altitude is known. Four satellites provide complete three-dimensional positioning.

calculating how long the satellite's radio signal took to reach the GPS receiver. As radio signals travel at the speed of light, very accurate clocks are needed to measure the small delay between transmission and reception. GPS receivers calculate the radio signal travel time by simultaneously generating the same pseudo-random code (PRC) that is transmitted by each satellite, and then backtracking to see when it generated the same PRC segment that was just received from the satellite. The radio signal of each satellite is identified by its unique PRC. The PRC is central to the functioning of GPS; without it, NAVSTAR radio signals would have to be much more powerful and we, as users, would still need television satellite dish antennae in order to receive the signals. As it is, GPS radio signals are so faint that they don't register above the earth's inherent background radio noise. Yet by using the satellite's unique PRC and some clever mathematics based on information theory, a GPS receiver fitted with a small antenna is able to extract the satellite's ranging signal from the background noise.

Because the GPS radio signal is so weak, locating a satellite's signal within the background noise is much more difficult than staying locked

Table 7.3
Sources and Sizes of Error When Using GPS to Establish Field Site Locations

Satellite clock error	0.6m
Ephemeris error	0.6m
Receiver errors	1.2m
Atmospheric errors	3.6m
Worst case S/A	7.5m
Total root-mean square	4.5–9m
Actual error is calculated by multiplying the total error above by the expected Geometric Dilution of Precision (GDOP) ranging from 4–6.	
Typical actual error	18–30m
If S/A is implemented	105m

SOURCE: Adapted from Hurn 1989.
NOTE: S/A is "selective availability," which is the error intentionally introduced by the U.S. Department of Defense to degrade the accuracy of the ground positioning system for nonmilitary users.

onto it. In fact, once a GPS receiver locks onto a satellite's radio transmission, the signal-to-noise ratio improves, and thus the power of that signal is effectively increased. This increase in signal-to-noise ratio becomes important when working in areas where the signal strength is unusually low, for example, under a forest canopy (Hurn 1989).

The GPS system uses the pseudo-random code to minimize antenna size and to provide the U.S. Department of Defense (DOD) with the ability to regulate access to the system and to control the locational accuracy available to the public. There are two separate forms of pseudo-random code: C/A and P. P code is encrypted, so only authorized military users have access to it. The DOD can degrade the C/A code available to the general public using an operational mode called selective availability (S/A). S/A essentially introduces an artificial clock error in a satellite's pseudo-random code, thus degrading the locational accuracy of autonomous GPS receivers. At the present time the accuracy of a GPS receiver is determined by the sum of all sources of error (18–30m), and can be degraded further by the DOD by implementing S/A (105m). (See table 7.3.)

Line-of-sight radio contact with three or four satellites is needed in order to determine location. GPS receivers must be designed either to listen to all four (or more) satellites simultaneously (this requires a separate channel designated to each satellite) or to listen to each of the four or more satellites sequentially (sequencing or multiplexing receivers can have as few as one channel). Single-channel receivers are the least costly to make, and they consume less power than do multichannel receivers. Yet as single-channel receivers must constantly switch from satellite to satellite, they spend most of their time extracting radio signals from the background noise and thus do not reap the signal-to-noise ratio benefits of multichannel receivers that can permanently lock onto a satellite's signal. The more channels a GPS receiver has, the longer an individual satellite can be locked onto, and the greater the signal-to-noise improvement. In addition, the more satellites a receiver can listen to, the more choice one has to select those satellites that provide the best geometry and the highest signal-to-noise ratio—both of which are important in determining the user's position. For this reason a three- or six-channel receiver is likely to perform better than a single-channel receiver when the line-of-sight to satellites is restricted and thus when signal-to-noise ratio is at its worst (e.g., under a forest canopy). Forest canopy can, of course, degrade or obscure the satellites' radio transmissions such that it is impossible for the GPS receiver to extract the signals from the background noise regardless of how many channels the GPS receivers have available. Areas where canopy closure exceeds 30% and the visual horizon averages more than 50° may preclude acquisition of the three- or four-satellite constellation necessary for location determination. Similarly, satellites higher than 70° above the horizon do not have the appropriate geometry for triangulation; thus even if visible, they cannot be used for position determination (Wilkie 1989).

In 1987 GPS receivers were not particularly attractive for the kind of field work discussed here. A single-channel variety cost about $16,000 and weighed a hefty 10kg. But technological advance has been stunning. Now hand-held three- and five-channel models that run on AA batteries can be purchased from marine electronics equipment stores for between $400 and $3,000.

Aerial videography Low altitude aerial videography with a GPS receiver (Sidle et al. 1992; Everitt et al. 1993; Bobbe et al. 1993) permits variable resolution imagery transects (oriented systematically or randomly) to be flown over areas that require field site characterization. Using a vertically mounted videocamera linked through a video character generator to a GPS receiver, the exact (within 100m, with selective availability) latitude and longitude of the terrain within the field of view of the

Table 7.4
Example of an Aerial Videography System for Field Surveys

Video	
Camera	Canon UCS5 Hi8
VCR	Sony EV-C100
Caption	Compix LP701 Compix Inc: 1-503-639-8496 or
SMPTE	Horita FP-50 GPS2
Capture	ComputerEyes/RT 24 bit color Digital Vision: 1-617-329-5400
GPS	Magellan Pro Mark V Magellan: 1-909-394-5000 Trimble Pathfinder or Ensign XL Trimble: 1-800-959-9567

NOTE: The Compix LP701 is a video caption system designed by Compix Inc. to document when and where a videotaped image was made. The LP701 accepts navigation data from the RS-232 or RS-422 output port of a variety of commercially available GPS receivers.

camera is written directly onto each video frame. Assuming that the camera operator is familiar with the survey area, the audio channel on the video tape can be used to describe in a systematic way terrain features of interest as they appear in the video camera viewfinder. Audio annotation of the video survey creates a permanent record of terrain characteristics—a record that proves vital for labeling digital image spectral classes or for generating training sites for a supervised image classification. The GPS receiver can also be used to locate (or relocate) and follow random or systematic transects within the study area. Table 7.4 provides an example of an aerial videography system.

Once a video survey is completed, individual sites of known geographic location and known land-cover type can be extracted from the video transects using a video frame grabber (Everitt et al. 1993). A video frame grabber captures a video image displayed on a computer screen; some loss of spectral information is likely with color frame grabbers because of the arithmetic rounding errors associated with generating and then decomposing an NTSC composite color video signal. The video information can then be stored on a computer, as images ready for digital analysis. These video-frame sample sites and associated audio terrain descriptions can serve as the basic ground-truth data to generate accurate land-cover classifications from satellite remote sensing imagery. Aerial videography provides a rapid, relatively inexpensive way of gathering detailed field data over relatively large areas. It is particularly valuable for mapping isolated or inaccessible regions.

Fine-Tuning the Training Sets

The accuracy of supervised classification depends on how typical the pixels within a training set are of the landscape feature they are purported to characterize. Accuracy also depends on sufficient separability of the spectral signatures extracted from each training set of pixels. Both factors are determined statistically, and their validity is a function of the size and representativeness of the training sets, and of the independence of pixels within each training set.

Determining how many pixels need to be included in a training set is not a simple matter of more being better. The cost of gathering ancillary information is also a concern. The minimum number of pixels to statistically estimate a spectral signature increases with the number of spectral bands (variables) involved. A rough guide is that at least $50x$ pixels should be sampled for each spectral-landscape class, where x is the number of spectral bands to be used to estimate the class signatures. For this minimum number of pixels to yield valid estimates, the value of each pixel must be independent of all others. In remote sensing imagery this is never the case, because adjacent pixels are more likely to be of the same class (open water, or forest, or field) than would be expected if the class of each pixel were determined randomly.

Adjacent pixels tend to be spatially autocorrelated. Autocorrelation deprives a training set of the necessary statistical independence of pixels. To overcome the spatial autocorrelation inherent within satellite imagery, we simply compute the spectral signature estimates using a randomly selected sample (10%) of the pixels within each training set. By subsampling the training set to obtain a statistically independent set of pixels, one can discover how much to increase the sample size of the original training set in order to achieve the degree of internal independence sought. For example, a 20% random sample would require a minimum of $50 \times 4 \times 1/0.20$ pixels within each landscape type training set for a four-band Landsat MSS classification. Other subsampling methods have been suggested that make use of correlograms (Cliff and Ord 1973) to determine the interpixel distance at which to systematically select a non-correlated pixel subsample (Labovitz and Matsuoko 1984). It is by no means clear, however, that the considerably more time-consuming correlogram method is any more effective than is a simple random sampling strategy, though it may result in a greater proportion of training set pixels being included in the signature extraction analysis.

We should point out that the spectral response of only common landscape features can be identified using these guidelines. Rare cover types may not occupy enough pure pixels to meet the criterion of $50x$ per subsample. If, for example, rare cover types constitute critical habitat for an endangered species, a supervised classification may be difficult to complete, thus forcing the use of unsupervised cluster analyses.

Once we have a sufficiently large *independent* sample of pixels, we need to determine if all pixels are, in fact, representative of the training set as a whole. The most effective way to do this is to conduct for each training set a hierarchical cluster analysis (an unsupervised classification, discussed at the beginning of this chapter). Visual examination of the resulting dendrogram will show which pixels are the last to be merged into the major cluster and can thus be considered atypical (*outliers*) and be excluded from the sample. Before deleting outlier pixels, however, the researcher should confirm if possible that they do not represent an important and unique feature.

Spectral signatures extracted from training sets are statistical representations of the characteristics of a particular class. In general, the mean, standard deviation, variance, minimum, and maximum for each spectral band, along with the variance-covariance matrix, are computed for all training sets. These univariate and bivariate statistics constitute the signatures for each class. The components of a class signature used for classification actually differ depending on the classifier to be used: maxima and minima brightness values for parallelepiped; class mean vectors for minimum distance; and both class mean vectors and covariance matrixes for maximum likelihood.

Once signatures have been generated from each training set, it is time to make two decisions. Are some classes not separable, given the spectral bands (or band transformations available), and do they thus need to be combined? What minimum combination of spectral bands offers the best discrimination between classes? To make these decisions, one begins by

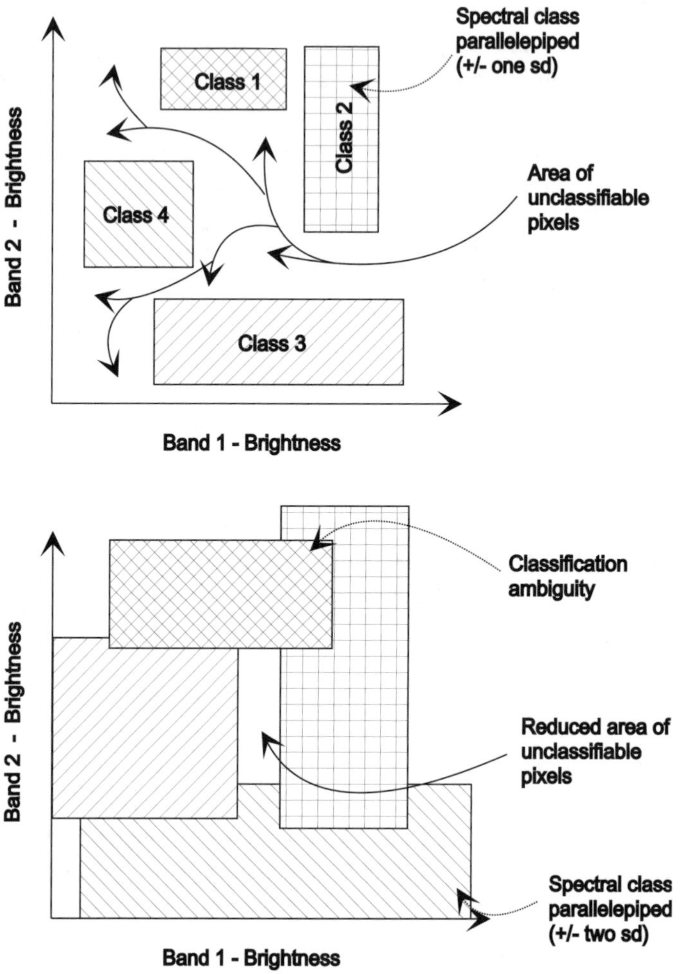

Figure 7.11 Parallelepiped classification and the trade-off between unclassified pixels and ambiguous class pixels. *Top:* the parallelepiped classifier can be applied to ensure that class boundaries do not overlap—thus pixels can be assigned unambiguously to one class. *Bottom:* alternatively, the method can ensure that few pixels remain unclassified. But parallelepiped classification typically cannot achieve both results at the same time. (This example uses two spectral bands.)

displaying on screen the shape and separation of signatures. Each signature can be represented by a two- or three-dimensional box (parallelepiped), the size of which is determined by the spectral range of the signature in the two or three bands that comprise the axes (figure 7.11). By displaying all signatures together in two-, three-, or multi-band combinations, we can determine how separable the classes are, and which band combination produces the greatest class discrimination.

If the box dimensions are based on maximum and minimum brightness values, then any overlap of signatures will mean that the parallelepiped classifier will be unable to discriminate between the overlapping classes and would thus arbitrarily assign pixels to either class if a pixel falls within the overlap zone. Max-min box plots very effectively display how well a parallelepiped classifier will work, and will indicate overlapping classes that should be merged. To select the spectral bands that best separate the training sets from one another, the spectral response of each training set within all spectral bands should be compared. Visual examination of the separability of training set signatures is both easy and informative (table 7.5, figure 7.12). There are, however, statistical techniques for determining both the spectral separability of signatures and the best combination of spectral bands to discriminate between training set signatures. (For a more detailed treatment of training set divergence measures, see Mather 1987:319–324 and Richards 1986:206–214.)

A rough assessment of how successful the training areas will be in generating discrete classes can be made by classifying only the pixels within the training areas. Training areas were chosen because each represented a distinct and relatively homogeneous example of a landscape cover type. But the remotely sensed imagery can be accurately classified into those land-cover types only if the spectral signatures do not excessively overlap. An ordination technique named *canonical variate analysis* (Hobbs et al. 1989) can offer help. It computes the linear combination of bands that maximizes the ratio of between-signature variation to within-signature variation, thus identifying which spectral bands are most effective at separating the spectral signatures of land-cover training sets. This method is also used to reduce the number of bands and eliminate correlation problems in a multispectral image whilst retaining the original information content.

Table 7.5
Descriptive Statistics of Training Set Pixels

	Band	Minimum	Maximum	Mean	Variance/Covariance Matrix						
Training Set					1	2	3	4	5	6	7
Water	1	64	73	67.97	2.99	0.84	0.67	0.39	0.79	-1.47	0.49
	2	20	25	22.12	0.84	0.97	0.53	1.10	0.81	-0.27	0.43
	3	15	20	18.20	0.67	0.53	0.83	0.93	0.76	-0.23	0.48
	4	13	28	16.35	0.39	1.10	0.93	3.99	1.80	0.99	1.05
	5	4	14	8.28	0.79	0.81	0.76	1.80	2.06	-0.02	0.75
	6	120	127	122.23	-1.47	-0.27	-0.23	0.99	-0.02	1.76	-0.02
	7	1	33	3.28	0.49	0.43	0.48	1.05	0.75	-0.02	2.69
Forest	1	3	75	68.70	6.49	0.59	0.77	-1.02	-0.69	-0.13	0.19
	2	24	30	26.77	0.59	1.33	0.70	4.89	4.91	0.19	1.36
	3	11	25	20.75	0.77	0.70	1.22	0.46	1.66	0.10	0.74
	4	67	116	94.22	-1.02	4.89	0.46	101.31	76.13	2.28	15.61
	5	40	77	59.18	-0.69	4.91	1.66	76.13	73.67	3.90	16.73
	6	126	131	128.17	-0.13	0.19	0.10	2.28	3.90	1.28	1.01
	7	9	22	15.82	0.19	1.36	0.74	15.61	16.73	1.01	4.83
Cleared	1	71	79	73.37	2.51	0.84	1.68	-5.83	1.96	0.09	1.80
	2	31	35	32.42	0.84	1.12	1.14	-2.71	1.85	0.37	1.10
	3	23	31	25.50	1.68	1.14	2.42	-6.11	2.91	0.55	2.31
	4	104	136	125.63	-5.83	-2.71	-6.11	36.08	-1.83	-2.22	-5.39
	5	80	97	90.03	1.96	1.85	2.91	-1.83	12.19	0.98	4.80
	6	127	130	128.76	0.09	0.37	0.55	-2.22	0.98	0.78	0.58
	7	22	30	25.39	1.80	1.10	2.31	-5.39	4.80	0.58	3.60
Urban	1	32	184	86.73	88.80	48.72	72.28	-47.60	38.14	3.18	49.35
	2	29	95	37.74	48.72	30.37	44.22	-23.55	27.01	1.32	30.87
	3	17	110	37.53	72.28	44.22	68.08	-46.51	39.75	3.05	47.42
	4	48	124	86.78	-47.60	-23.55	-46.51	152.35	26.94	-8.66	-26.01
	5	53	143	82.06	38.14	27.01	39.75	26.94	82.40	1.11	42.20
	6	134	147	139.95	3.18	1.32	3.05	-8.66	1.11	3.67	2.78
	7	20	72	32.67	49.35	30.87	47.42	-26.01	42.20	2.78	40.85

NOTE: Data in this example are drawn from the pixels bounded by white polygons shown in figure 7.7.

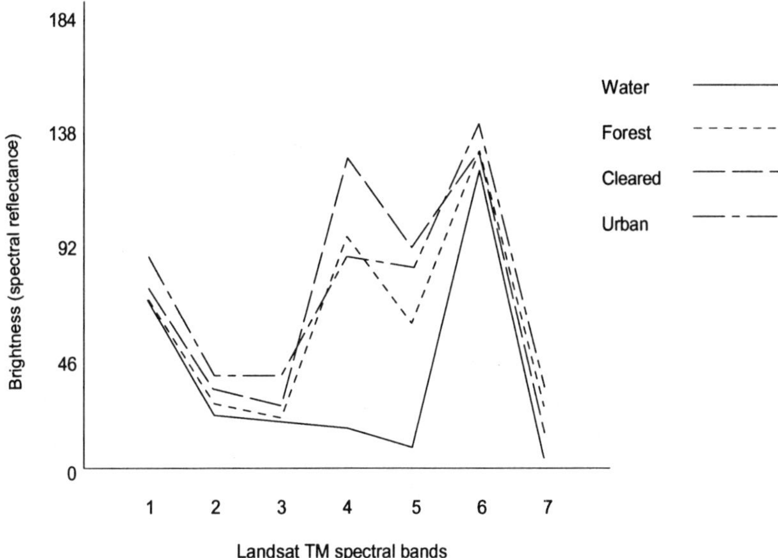

Figure 7.12 Using brightness profiles of training sets to select the best wavebands for supervised classification. Displaying the brightness of each training set (water, forest, cleared, and urban) in all spectral bands helps one select those bands that best discriminate among the features of interest. In this example, bands 4 and 5 appear to offer the best opportunity for classifying pixels into the four land-cover types.

Selecting and Using a Classifier

Now that a separable set of spectral signatures has been generated and the best combination of spectral bands to discriminate between signatures has been identified, we can proceed with selecting and applying a classifier to the imagery. Three techniques are used commonly as classifiers: parallelepiped, minimum distance to centroid, and maximum likelihood. These three classifiers differ in the quantity of training data needed, the speed of the analysis, and the average classification error expected.

Parallelepiped Classification

This classifier is based on simple Boolean logic and does not require a pixel to be algebraically compared to individual signatures in order to assign them to a specific class. As a result, this is the fastest of the three classifiers.

Let us assume, for example, that p spectral bands were determined to best discriminate among the training set signatures. For each signature we make the assumption that the distribution of brightness values can be represented by a p-dimensional parallelepiped. The boundaries of each parallelepiped are determined either by the maximum and minimum or by the mean ($\pm$ one standard deviation) of the brightness values of pixels contained within each training set.

The image is classified by examining the brightness values of each pixel in all p dimensions to see whether it falls within the boundaries of any of the parallelepipeds, and labeling it accordingly. As it is very unlikely that the number, size, and orientation of all parallelepipeds are such that they cover the whole p-dimensional space, some pixels will not fall within the boundaries of any parallelepiped and cannot, therefore, be classified. Unclassified pixels are usually assigned a label of zero. If parallelepiped boundaries were determined using the mean ($\pm$ one standard deviation) method, then the size of each parallelepiped could be increased by using a larger number of standard deviations from the mean. Though this may reduce the area of unclassified pixel space, it generally results in overlap of individual parallelepipeds. Overlap introduces the new problem of pixels lying within the boundaries of two or more parallelepipeds (figure 7.11). When parallelepipeds do not define a unique p-dimensional space, there is no logical rule to assign pixels to one or another overlapping parallelepiped. In most software used for remote sensing analysis, ambiguous class pixels are assigned to the first encompassing parallelepiped encountered. In this case, the order in which parallelepiped membership is evaluated will have a strong influence on the final classification.

The advantages of parallelepiped classification are that it is quick and requires very little information about the spectral signatures of training sets (maximum, minimum, and mean brightness). But because this method assumes a box-shaped distribution of brightness values, it is beset with the problems of either too many unclassified pixels or pixels with multiple class membership.

Minimum Distance to Centroid Classification

The problems of unclassified or ambiguously classified pixels can be avoided by using the minimum distance method. Like parallelepiped classification, this classifier uses the mean brightness values for each spectral band, as computed from training set pixels.

Figure 7.13 Class boundaries typical of a minimum distance to centroid classifier. All pixels can thus be unambiguously assigned to one class, and no pixels will remain unclassified.

The first step in the minimum distance form of classification is calculation of a set of mean vectors that constitute the center location (centroid) of each spectral signature in *p*-dimensional brightness space. For each pixel, the classifier calculates the Euclidean distance to each class centroid,

$$Pixel{\rightarrow}centroid_k = \sqrt{\sum_{b=1}^{p}(BV_{rcb} - \mu_{kb})^2}$$

where p is the number of bands, r is the row and c the column location of the pixel, k is the class, b is the spectral band, BV_{rcb} is the brightness value of pixel rc in band b, and μ_{kb} is the mean of class k in band b.

After the distance to each class centroid is calculated for the pixel in question, the classifier assigns that pixel to the closest class. The logic of this classifier allows all pixels to be assigned unambiguously to one class. This is true because, unlike regular-sided parallelepipeds, the *p*-dimensional polygons that surround each centroid are irregularly shaped and their boundaries are contiguous, leaving no empty and thus unclassified space (figure 7.13).

Though it may seem important to be able to unambiguously classify all pixels within an image, if the classification scheme is not exhaustive or if training sets are not available for a few rare classes, then some pixels should indeed remain unclassified rather than including them erroneously within a class. As atypical pixels are likely to exist near the boundaries of centroid polygons, it is possible to set minimum distance thresholds to exclude such outlier pixels. For example, if the pixel-to-centroid distances were calculated as normalized deviation units, we could set the minimum distance threshold to not exceed 1.96 standard deviation units from the centroid. By doing so, we would expect to exclude about 5% of the pixels, because they lie at a distance beyond that specified by the threshold, assuming that the data are normally distributed. Furthermore, the minimum distance classifier either can compare raw distances to determine class membership or can normalize distances by the standard deviation of each band. The latter is usually more effective when the number of pixels within training sets is large. In contrast, the simple raw distance measure may produce a more accurate classification when training sets are small and thus contain few pixels.

Minimum distance classification overcomes the unclassified and ambiguously classified pixel problems of the parallelepiped classifier, without requiring additional training set information. Its major drawback is that by relying solely on mean vectors extracted from the training sets, it assumes that the class members are distributed symmetrically around the centroid.

Maximum Likelihood Classification

If we can establish how class members are distributed around the centroid, we can more accurately assign membership of a pixel to a specific class, and thus minimize the error associated with image classification. Rather than assuming that the p-dimensional distribution of a given class is box shaped (parallelepiped) or spherical (minimum distance), the maximum likelihood method uses spectral band covariances of the training sets to determine the orientation and relative elongation of the p-dimensional probability distribution around the mean of each class. Maximum likelihood classification determines the true shape of the distribution of each class, and as a result is considered the most accurate classifier.

Maximum likelihood classification is based on a fundamental theorem of probability established by an English clergyman, Thomas Bayes. The a posteriori probability, $P(i|x)$, is the likelihood that a pixel with brightness value vector x composed of p elements (bands) is a member of class i. This membership probability is calculated for each class from the determinant of the class variance-covariance matrix, and the Mahalanobis distance (the pixel-to-class centroid distance corrected for the variance and covariance of the class i), using the equation

$$P(i|x) = 2\pi^{-p/2}|S_i|^{-1/2} Exp\ [-(Mahalanobis_i)/2]\ ,$$

where S_i is the variance-covariance matrix for class *i*.

Once the function $P(i|x)$ is calculated for each class, then the pixel is assigned to the class for which $P(i|x)$ is largest—that is, the class for which the pixel has the greatest probability of being a member. This equation assumes that each class occupies the same total area within the image, and thus that each should be represented as having an equal probability of occurring. In most landscapes covered by satellite imagery, this assumption will not be true. However, rarely do we know a priori what the proportional composition of each class is within the image. In fact, most classifications must be completed before one can determine the relative coverage of each class within an image. If we can use ancillary information, such as altitude or soils, to determine which areas within the image are likely to be represented by which class, then we can estimate the a priori occurrence probability of specific classes within the image. This information can be incorporated into the maximum likelihood classifier,

$$P(i|x)refined = P(i|x)P(i)$$

where $P(i)$ is the a priori probability of class *i* occurring within the image. Inclusion of both a posteriori and a priori probabilities into the classifier function will improve the accuracy of the classification (Strahler et al. 1980).

Plate 9 depicts the results of applying parallelepiped, minimum distance, and maximum likelihood classifiers to bands 2, 4, 5 and bands 4, 5 from a window of the Landsat TM image of Calais, Maine.

Improving the Accuracy of the Classification

The next step after completing the classification is to improve its accuracy. To improve the accuracy, one must first determine it.

Determining the Accuracy of the Classification

If the parallelepiped classifier was used, or if the minimum distance or maximum likelihood classifiers were used with a threshold set, then unclassified (or zero class) pixels will exist within the classified image. If the number of zero pixels is large (exceeds 10%), then we might question how exhaustive the classification scheme is. We also may come to suspect that the training sets upon which the classification is based may not after all be representative of the terrain.

If unclassified pixels are not a problem, then we can move on to establishing both the overall accuracy of the classification and the location-specific accuracy. To do this, we need information about the true composition of a sufficient number of locations, which we can then compare with those labeled in the classified image (our created map). Ancillary information about the true composition of specific areas within the image can be obtained from published maps, aerial photography or videography, and field visits.

As a rough rule of thumb, we should attempt to describe the true land-cover type of at least 10% of each class within the image. In this way we can effectively evaluate the accuracy of our classification. A more formal strategy for determining the number, *N*, of ground-truth locations (pixels) to sample is

$$N = \frac{4(a)(100-a)}{e^2},$$

where a is the accuracy we hope to achieve in our classification, and e is the amount of error we are willing to accept in assessing the classification accuracy—given the number of ground-truth points sampled. If, for example, we are looking for 90% accuracy ($a = 90$) with 95% confidence ($e = 5$), then we calculate that at least 144 independent ground-truth points need to be identified for each class. Less accuracy and less confidence in the analysis means fewer control points are needed. As with training set field sites, the location of ground-control or classification verification sites should be distributed in a stratified (by class) random pattern throughout the image.

As with signature extraction training sets, the minimum area of an individual ground-truth site should be related to the spatial resolution of the sensor and the accuracy of the geometric correction. That minimum area can be calculated using

$$Area = (m(1 + 2p))^2,$$

where m is the spatial resolution in meters, and p is the geometric accuracy in pixel units (Justice and Townshend 1981).

For an overall assessment of classified image accuracy, we simply take the proportional composition of true land-cover types as determined from ground-truth data collection and then compare this figure to the composition shown within the classified imagery. To assess the location-specific accuracy of the classification, we need to compare the class of a set of pixels with their corresponding cover-type on the ground. To do this, the location of ground-truth points must be established using the same coordinate system as that of the classified image, and at an accuracy that permits the corresponding pixel to be identified unambiguously. The easiest way to accomplish this is to create an image of the same scale and geographic coordinate system using the ground-truth information. It is then a very simple matter for the computer to overlay the classified and ground-truth images and make pixel-to-pixel comparisons. A ground-truth image can be created either by digitizing ground-truth information from maps or aerial photographs or by creating a vector feature file from GPS-located field survey data. Most image processing software can read vector files composed of the class type, and *x*,*y* coordinate locations of point or polygon features. These vector images can then be converted to a raster (cell or pixel) format of the same scale and coordinate system as that of the classified image.

Table 7.6 shows results of a pixel-by-pixel comparison. The data are presented in a square matrix where the number of rows and columns equals the land-cover classes used in the study. Land-cover classes in the ground-truth image head the rows, and the same classes for the classified

Table 7.6
Image Classification Error Matrix

	sample % of total pixels	Classified Image Pixels				
		Water	Forest	Regrowth	Urban	Row Total
Ground Truth Image Pixels (known land-cover types)	Water 10%	180	0	0	9	189
	Forest 5%	0	452	89	2	543
	Regrowth 16%	0	123	221	34	378
	Urban 3%	0	25	8	97	130
	Column Total	180	600	318	142	1240

NOTE: Image classification error statistics are as follows:
percent correct = 76.61%
Kappa = 0.65 (Rosenfield and Fitzpatrick-Lins 1986:224)
AMI = 833.26 (Ulanowicz 1986:81–95)

image head the columns. Matrix element e_{ij} is the number of pixels labeled class *i* in the ground truth image and class *j* in the classified image. For example, e_{14} is the number of "water" pixels in the ground-truth image that were classified "urban."

The main diagonal elements of the matrix (e_{ii}) are the number of pixels that are labeled the same in both images. The matrix can be considered an *error matrix, E,* with the off-diagonal elements (e_{ij}) representing the errors of omission (pixels that from ground truth data we know are class $_i$ but were labeled as class $_j$ in the classified image) and commission (pixels that were labeled as class $_i$ in the classified image but were actually members of another class according to the ground truth data) that occurred when we assigned pixels to specific land-cover types (Finn 1988).

Several statistical techniques have been developed by which the error matrix can be used to determine the accuracy of the classified image. The three most common techniques are *percent correct, a coefficient of agreement Kappa* (Rosenfield and Fitzpatrick-Lins 1986; Cohen 1960), and *average mutual information* (Ulanowicz 1986). Percent correct is the ratio of the sum of the "correct" diagonal elements of the matrix e_{ii} to the total number of pixels used in the accuracy assessment. The percent correct technique only makes use of the pixels that were labeled accurately, and does not separate errors of commission and omission.

$$\textit{Total \% Correct} = \frac{\sum_{i=1}^{c} e_{ii}}{N} \qquad \textit{Class \% Correct} = \frac{e_{ii}}{\sum_{j=1}^{c} e_{ij}}$$

Kappa measures the agreement, beyond chance, between two maps, taking into account all elements of *E*. To calculate Kappa, each element of *E* is divided by the total number of pixels (*N*) to produce a matrix (*P*). The elements in matrix *P* are the probabilities, $p_{ij} = p(y_j, x_i)$, of a pixel being *i* in image *X* and *j* in image *Y*. Percent correct is calculated in *P* as

$$\frac{\sum_{i=1}^{c} p_{ii}}{1}.$$

The *i*th row sum of *P* is the probability $p(x_i)$ of class *i* occurring on map X. Similarly, the *j*th column sum is the probability $p(y_j)$ of class *j* occurring on map *Y*. The probability of a pixel being *i* in X and *i* in *Y* by chance is $p(x_i) * p(_{yi})$. Therefore, if we remove chance agreement from the percent correct technique, we get Kappa given by

$$\hat{K} = \frac{\sum_{i=1}^{c} p_{ii} - \sum_{i,j=1}^{c} [p(x_i)p(y_j)]}{1 - \sum_{i,j=1}^{c} [p(x_i)p(y_j)]}$$

Average mutual information (AMI) is a general index from information theory (Ulanowicz 1986) that can be interpreted in this case as the quantity of information shared by the two images, *X* and *Y*. Succinctly, it is the reduction in storage requirements for image *X* if image *Y* is known. If log base 2 is used, the information content (or uncertainty) of image *X* in bits (without knowing *Y*) is given by

$$H_x = -N \sum_{i=1}^{c} p(x_i) \log_2 p(x_i)$$

where H_x is the minimum number of bits required to store the information contained within image *X*. The conditional probability, $p(x_i|y_j)$, that class *i* will occur on map *X* at the same pixel that class *j* occurs in map *Y* is calculated using

$$p(x_i|y_j) = \frac{p(y_j,x_i)}{p(y_j)}.$$

Average mutual information is then computed from

$$AMI = N \sum_{i=1}^{c} \sum_{j=1}^{c} p(y_j,x_i) \log_2[p(x_i|y_j)/p(x_i)].$$

Finn (1993) showed that the AMI technique indicated consistency more than correctness; it should therefore be used in combination with % correct or Kappa.

Improving Classification Accuracy

If the classifier accuracy has been judged less than acceptable, the next step is to make improvements. Classification accuracy is (paradoxically) inversely related to the spatial resolution of the sensor, because as spatial resolution increases so too does landscape heterogeneity and the number of unique pixel brightness values. As a result, and assuming a constant error probability, as the number of unique pixel values increases, so too will the number of pixels assigned to the wrong class. Similarly, the number of classes is inversely related to classification accuracy. This is less surprising, as a single-class classifier would always be 100% accurate.

How do we make improvements? There are several approaches. If the accuracy of a classification is poor we can

- modify the resolution of the image, using a smoothing filter (smoothing is also often used to remove the salt-and-pepper appearance of the classified image that occurs because individual pixels of one class are often embedded as islands in another);
- reduce the number of classes by merging those that are most similar;
- attempt to provide additional information to the classifier by increasing the size of training sets;
- obtain more ground-truth information; and
- add ancillary information (e.g., elevation, slope, and aspect) to improve our understanding of how the terrain interacts with and reflects radiation.

Detecting and Monitoring Temporal Change in a Landscape

Remote sensing image analysis not only provides us with the ability to create thematic maps that represent accurately the type, shape, and distribution of landscapes that characterize a given area; it also offers the opportunity to monitor areas over time to determine if and how landscape composition changes. Visual interpretation and comparison of multidate aerial photographs have been used very successfully to detect landscape changes over time. For example, the National Wetlands Inventory (USFWS) has provided a very detailed record of the changes in wetland extent and location, but the process is slow, labor intensive, and costly. These disadvantages of visual interpretation are what make computer-assisted methods attractive.

Computer-assisted methods of remote sensing image analysis can allow us to detect changes in landscapes very quickly, over very large areas, and as frequently as imagery can be obtained. Our ability to track rapidly changing events—such as floods, rainfall pattern, and crop maturation—and to monitor long-term changes wrought by successional processes, human land use, and climate, over large areas, is a much more practical proposition since the launch of earth resource sensing satellites. Unlike aerial photography, satellite remote sensing imagery is particularly suited to detecting changes at regional scales. One reason is that a satellite revisits the same location on the earth's surface on a regular schedule (allowing change to be detected over an hourly, daily, or longer period). Revisits are made at the same time of day (thus avoiding differences in sun angle from one time period to the next). Satellites can also be depended on to record spectral data consistently and at the same scale. Finally, satellite imagery is attractive because it is relatively inexpensive to purchase and process, for regional scale monitoring.

Interpretation of remote sensing imagery depends on our ability to visually or digitally distinguish the spectral reflectance of relevant landscape features. As we have discussed, it is very important, therefore, to select imagery during a time of year when the features to be identified exhibit the greatest spectral contrast with the background. This dictum is equally important for change detection, because success is predicated on whether the change phenomenon that we hope to measure manifests a detectable difference in the spectral reflectance of pixels from one time period to the next. This raises the question whether changes in atmospheric conditions, sun angle, phenology, soil moisture, lake level, and so on between $time_1$ and $time_2$ are part of the phenomenon that we wish to measure, or constitute noise that will obscure the detection of real change.

If seasonal changes are not of interest, imagery for $time_2$ should be obtained as close as possible to the anniversary date of the acquisition of the $time_1$ image. This, of course, assumes that weather and vegetation growth patterns are invariate from year to year. Weather patterns are seldom accommodating of our interests, but use of anniversary images does help to minimize differences in reflectance attributable to seasonal factors. If data are available, choosing an acquisition date for the change detection image that has the same number of growing degree days as the original image may provide better results than using a simple anniversary date selection criterion.

Coregistration of Change Detection Images

For change detection to be successful, the $time_1$ and $time_2$ satellite images must be spatially coregistered. This means that the displacement between a pixel in image $time_1$ and its homologous pixel in the $time_2$ image does not exceed half the spatial resolution of the sensor system (15m for Landsat TM imagery).

Even when images are obtained from the same satellite system, such as Landsat 4 or SPOT 1, slight changes in the attitude of the satellite (its pitch, roll, and yaw) may produce a spatial offset in the two images: a pixel at row *i* and column *j* in the $time_1$ image may not be located over exactly the same area as its homologue in the $time_2$ image. Both images that are to be used for change detection should therefore be coregistered, either by using the image-to-image technique described in chapter 5 or (preferably) by geographically correcting each image to a standard map coordinate system. As exact pixel-to-pixel registration is exceedingly difficult to achieve, noise (error) is introduced in the images. It is often necessary, therefore, to first smooth the images by processing with a low-frequency filter.

Strategies for Detecting Change

Four primary strategies are used for detecting change between two images: (1) single-band raw-image algebra, (2) spectral vector analysis, (3) principal components analysis, and (4) classified image comparison. The first strategy is used for comparing two temporally different images of a single spectral band. The other three can be applied to multispectral imagery.

Single-band raw-image algebra This strategy for detecting change involves either *image differencing* (which is the simple subtraction of one image from another obtained at a later date) or *image ratioing* (which is the division of one image by another obtained at a later date). Both techniques are applied to the same single band (channel) in both images. To minimize brightness value differences attributable to atmospheric effects, images for both dates should first be corrected for atmospheric errors (chapter 5).

Selection of the most appropriate band for image differencing or image ratioing requires that the user have some understanding of the spectral nature of the change phenomenon to be measured. For example, visible red wavelengths are best for detecting the outright loss of forest to urban development, but changes in the health of vegetation are best detected in the near-IR. Changes in water quality are most clearly seen at blue-green wavelengths. The user, through experience or trial and error, must choose the satellite image band with a spectral resolution that is best suited for detecting the change phenomenon of interest.

In image differencing, coregistered images of the same waveband are overlaid, and a new image is created by systematically subtracting the brightness value of each pixel at $time_1$ from that of its corresponding pixel at $time_2$. The new image is composed of zero values where no change in brightness occurred and positive and negative values in areas of change. Most landscape classes are composed of a range of spectral brightness. As members of each landscape class are likely to show some change in spectral reflectance over time (for example, trees grow and buildings are restored or replaced), there will be a range of values centered around zero that reflect areas that have changed but that have not become a different land-cover class (figure 7.14). The question is, therefore, what value indicates the boundary (threshold) between pixels that have not become different land-cover types to the degree we are interested in and those that have.

Detecting change in multiband imagery When, for example, an area of forest is cleared for agriculture or a suburban development, its spectral response is likely to change at several different wavelengths. Using simple Euclidean geometry, the distance and direction between the brightness levels of two spectral bands in before- and after-images will correspond to a spectral change vector (figure 7.15). If the vector exceeds a certain threshold, then one can assume that the land cover has changed to a degree that interests us. Direction of the vector tells us the cover type that the area has transformed into.

Determining appropriate threshold values for change and how vector direction is associated with cover type transformation is achieved preferably by examining the spectral vectors of areas that have undergone known changes, thus developing a series of change vector archetypes. This technique can be applied to any number of bands. However, determining change vector archetypes becomes much more complex as the number of bands increases.

Principal components analysis (PCA) can also be used to detect landscape change. Coregistered before- and after-imagery, each with N bands, are combined into a single $2N$-dimensional image, from which an equal number of principal components (see chapter 6) are computed. Careful

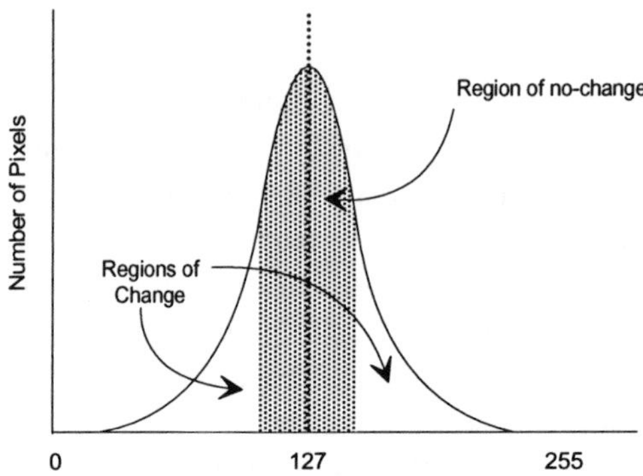

Figure 7.14 Determining a brightness threshold beyond which land-cover change will be deemed to have occurred. To detect landscape change that has occurred between the acquisition dates of two images, the images are coregistered and the brightness of each pixel in image 1 is subtracted from the brightness of each corresponding pixel in image 2. To ensure that the value of each pixel in the difference image lies within an 8 bit data range (0–255), pixel difference can be calculated using this equation: Pixel difference = (255 + pixel x_1,y_1 − pixel x_2,y_2) ÷ 2. Using this equation, very bright and very dark areas in the difference image will represent areas that undergo change. To avoid confusing the minor variations in brightness that occur within any given land-cover type with those associated with a change from one cover type to another, a difference threshold is established—usually in standard deviation units from the mean (127, if we assume that most areas have not changed).

examination of all components, when combined with a basic understanding of the type and location of change within the area, often reveals that one or more components highlight areas of change. These components can then be used as the basis for an unsupervised classification, which is likely to generate a set of change classes. Richards (1984) used PCA of multidate Landsat MSS imagery to locate burned and regenerating areas within southeastern Australia. In this case components 3 and 4 emphasized the areas of change, and were used to create a thematic change map using unsupervised classification.

To examine forest change in the Ituri region of northeastern Zaire, Justice et al. (1993) applied a change detection technique developed from pattern recognition studies (Tou and Gonzalez 1974) that had been used to detect landscape change in Mato Grosso, Brazil (Nelson et al. 1987). All six usable bands (band 1 was excluded because of haze) within 1976

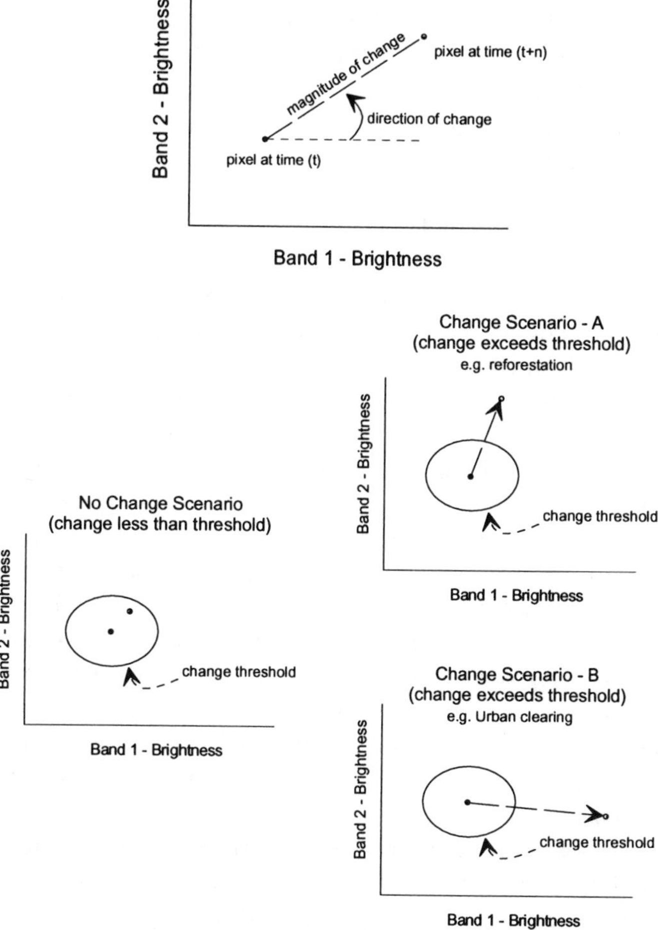

Figure 7.15 Detecting change in multispectral imagery. This example shows, using two spectral bands plotted in two dimensions, how the brightness of a pixel at time t can change in both spectral bands at time $t+n$. The direction of change and the Euclidean distance that describes the shift in pixel brightness between the two sampled times constitutes a vector of change. If that vector exceeds a threshold defined by the researcher, then significant change is considered to have occurred. Note in the lower right of both graphs how the direction of change, when combined with the researcher's knowledge of the spectral response of landscape features in different wavebands, indicates the type of change that has occurred.

and 1986 Landsat MSS scenes over the same area in northeastern Zaire were combined. The six-band dataset was used to generate fifty clusters that were then grouped visually into information classes. Information classes included those of no change (e.g., primary forest, secondary forest, nonforest) or change classes (e.g., forest to nonforest, nonforest to forest, nonforest to secondary forest).

One final commonly used method of conducting change analysis for multiband imagery is to undertake separately a land-cover classification of the before- and after-images and then conduct a pixel-by-pixel comparison of the two thematic images. The results are placed in a change matrix. Table 7.7 shows a change matrix generated by a pixel-to-pixel comparison of two land-cover classified images. Examining the matrix we can see that urban areas have increased substantially in the 23 years between image acquisitions (from 2,330 pixels in 1972 to 4,011 in 1995). Furthermore, we can see that the area of the urban landscape increased primarily through the conversion of regrowth forest (1,200 pixels of regrowth forest turned into urban pixels by 1995), although even water pixels were converted to urban pixels, as filling of flooded areas is a common way to increase the area of urban real estate throughout the world. The change matrix also shows an absolute decline in the area of forest, although 23 pixels that were regrowth forest in 1972 had returned to mature forest by 1995. All differences detected between classified images are considered "changes" and may be due to phenomena of interest or may not. The inaccuracy of this method is of course the combined inaccuracies of each individual classification.

The methods for detecting change in multiband imagery were supplemented in 1992 by Sader and Winne (1992), who developed a new technique for characterizing and monitoring change using multitemporal satellite remote sensing imagery. Rather than examine change between just two points in time, Sader and Winne tracked changes in forest cover in Maine across a three-image time series (1978, 1981, 1987). The three images were coregistered and an NDVI image computed for each scene. The three NDVI scenes were then displayed simultaneously by assigning individual images to one of the three colors in a standard computer display. In this case the 1978 image was assigned to blue, 1981 to red, and 1987 to green. Using an understanding of additive color theory and the fact that NDVI brightness is directly related to the biomass of vegetation, the authors were able to interpret forest change according to the color of

Table 7.7
Classified Image Landscape Change Matrix

		Classified Pixels in 1972 Image				
		Water	Forest	Regrowth	Urban	Row Total
Classified Pixels in 1995 Image	Water	2100	0	0	9	2109
	Forest	0	5546	23	0	5569
	Regrowth	0	1230	3428	0	4658
	Urban	240	250	1200	2321	4011
	Column Total	2340	7026	4651	2330	16347

a landscape patch as displayed on an RGB computer monitor. (This method will be explained in more detail in the next chapter.)

Generating Maps of the Analysis Results

During and after you have modified and analyzed remote sensing imagery to address a specific question or suite of questions, you will need to produce output of the results for review, publication, and dissemination. This section describes briefly the process for generating transparencies and prints of images. A detailed treatment of the process of generating high-quality cartographic output is given in the IDRISI for Windows Version 1.0 Manual (McEndry et al. 1995) and in the Map and Poster Layout Application Note published by MicroImages, Inc. (Skrdla 1992b).

Most image analysis and GIS software packages have the ability to output images and GIS data layers (vector representations of rivers, roads, elevation contours, etc.) directly to a printer or film recorder. However, exactly how an image will be reproduced depends on the technology that the printer or film recorder employs to create output. One must bear in mind that colors are generated very differently by color display devices, such as computer screens, and by color output devices.

As we noted earlier, computer screens use red, green, and blue (RGB) light at different intensities to create all the colors of the spectrum through a direct or *additive color process.* When red, green, and blue light are added at equal levels, the resulting hue varies from black (R, G, B =

0, 0, 0), through various shades of gray (e.g., R, G, B = 100, 100, 100), to white when all colors are set at full intensity—i.e., 255 for an 8 bit video processor (R, G, B = 255, 255, 255). Altering the relative proportion of the R, G, and B *additive primary colors* creates all the hues in the spectrum (e.g., deep purple = 179, 20, 193; gray yellow = 205, 199, 29).

Laser film writers and CRT imagers use the RGB additive color process to create high-resolution output of images suitable for publication. CRT imagers use a small, but very high resolution (>3,000 lines), video screen to display the image. A dedicated fixed-focus, 35mm camera is then used to capture the CRT image on print or slide film. These devices are a considerable improvement over the admittedly simple alternative of shooting print or slide film with a 35mm camera directly from the computer display screen in a darkened room. Shooting from a computer screen yields resolutions no better than that of the screen (640×480, 800×600, or 1,024×768). Photographing images directly off the computer screen works reasonably well when the shutter speed is set slower than the screen rewrite rate (i.e., <1/30 second) and the display monitor is of the new "flat screen" type. With older "curved screen" monitors the resulting slides or prints tend to suffer barrel distortion—bowing toward the edges.

Printers, in contrast to computer displays and film recorders, use dyes rather than light to create colors. Dyes generate color by absorbing certain wavelengths from a light source (e.g., the sun or incandescent light bulb), and thus only those wavelengths that are reflected by the dyes are visible to the viewer. In the case of printed color, white is created by the absence of dyes—i.e., all light from the source is reflected. Black is generated from equal proportions of cyan, magenta, and yellow (CMY), because these three *subtractive primary colors* absorb all wavelengths from the radiation source when they are combined. A mixture of cyan and yellow dyes reflects the color green, whereas magenta and yellow dyes produce the appearance of red. (In reality, the subtractive primary dyes—cyan, magenta, and yellow—do not filter out all light when combined in equal proportions. Thus printers often use a fourth dye—black—to produce a true, dense black: CMYK.)

Color display devices create color by using *additive primaries* combined directly by the radiation source and conveyed to the viewer as a hue. Printing devices create colors using *subtractive primaries* that absorb certain wavelengths from a radiation source, leaving the remaining reflected light to convey the hue to the viewer. As display and printing devices employ different methods for generating colors, the image that you see on the screen of your computer may not appear exactly the same when printed. Very likely the translation from RGB (*additive*) colors to CMY (*subtractive*) colors will not be perfect. Some software packages attempt to minimize variability in the color of displayed and printed images by using a standard color-matching system, such as Pantone for both devices.

Display devices are able to vary the amount (intensity) of each primary color, but printers are only able to apply the color or not. Printing devices cannot build layers of color, as does a painter, to create lighter or richer hues. Printers have overcome the problem of creating shades of gray (tonal levels) or ranges (values) of color by a process called *halftoning*. Halftoning generates different shades, or values, by printing dots of varying size within a standard grid pattern. As the number of points in the grid is constant, changing the size of each dot of ink will alter the proportion of white background visible. The resulting halftone, when viewed from a distance, is perceived by the eye not as dots of black or color on a white background but as a shade of gray or a more pastel value of the color. If you look closely at a newspaper image, you should be able to see the dots from which a photographic image is composed. Notice, too, that the larger the dots, the darker is the shade of gray that you perceive.

To print all the colors of the spectrum, a halftone screen is prepared for each of the three primary subtractive dyes (and black, if desired), with the grid alignment oriented at 45° for black, 15° for cyan, 75° for magenta, and 0° for yellow. By overlaying this set of four halftone grids, the color dots form a rosette pattern that is perceived by the eye as hues of different saturation and intensity. If the screens are not precisely aligned, the resulting moiré pattern causes the image to appear to vibrate, detracting severely from the quality of the final printed product. As many printers lay down a single color at a time, and therefore must make three or four passes to generate each image, paper alignment is critical to successful reproduction of color imagery.

Quality of the printed image is also related to the number of points in a halftone grid. This measurement is expressed in lines per inch (lpi). The higher the lpi, the greater is the density of dots and the sharper the resulting image. Newspapers are printed at a fairly coarse halftone resolution (70 lpi), whereas books and glossy magazines are usually printed at

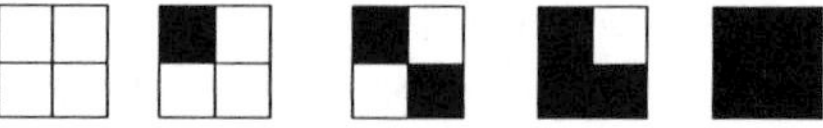

levels of grey (5) with 2 x 2 halftone cell

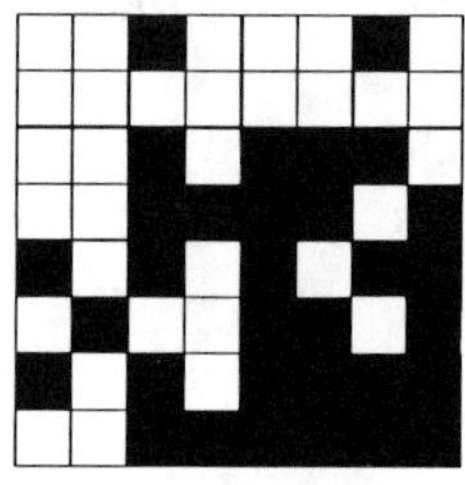

Figure 7.16 Halftone cells for dithering. *Top:* the five levels of gray possible from 2×2 halftone cells. *Bottom:* an image constructed from a total of sixteen 2×2 halftone cells.

150 lpi or higher. Unlike traditional lithographic printing where the size of halftone dots is variable, most digital printing devices can print dots of only one size.

To simulate variable dot size capabilities, halftone digital output devices use a process called *dithering* that groups variable numbers of dots of one size into blocks of *halftone cells.* The size of the halftone cell is equal to the total number of rows and columns of dots used to fill the cell (figure 7.16). By varying the number of halftone cells that contain ink dots, dithering can simulate the variable size (% black to white) of true halftoning. The dimensions of the halftone cells in the dithering process determine the number of shades of gray (tonal levels) that can be created (row × column + 1). However, as the halftone cell dimension increases (increasing tonal resolution), the spatial resolution declines. For example, a good quality laser printer has a resolution of 600 dpi (dots per inch). If we use an 8×8 dithering process to create 65 tonal levels (including white), then the effective halftone resolution of the printer is 1/8th of 600 lpi (if we consider a dot equivalent to a line)—i.e., 75 lpi. To translate

Table 7.8

Digital Output Devices for Producing Hard Copies of Remote Sensing Images

CMYK Device	**Technology**	**Color**	**Halftone**	**Resolution**	**Size**	**Quality**	**Purchase Price**
Dot matrix	Inked ribbon and impact pins	Yes	dithering	200dpi	11 × 14	✓	$150–$450
Inkjet	Liquid ink and spray nozzles	Yes	dithering	300dpi	11 × 14	✓✓	$500–$3,000
Thermal wax	Wax ribbon thermal pins	Yes	dithering	300dpi	11 × 14	✓✓✓	$1,500–$8,000
Dye sublimation	Dyes mixed when vaporized	Yes	variable	300dpi	11 × 14	✓✓✓✓	$1,500–$15,000
Phase change	Melted crayon and spray nozzles	Yes	variable	300dpi	11 × 14	✓✓✓✓	$6,000–$12,000
Laser	Black toner, electrostatic drum	No	dithering	600dpi	8 × 11	✓✓✓	$700–$5,000
Color laser	Colored powder, electrostatic drum	Yes	dithering	300dpi	8 × 11	✓✓✓	$5,000–$10,000
Electrostatic	same as laser	Yes	random	400dpi	44 × NA	✓✓✓✓✓	$25,000–$80,000
Lithography	photoresist metal plates	Yes	variable	3000dpi	34 × 36	✓✓✓✓✓	NA
RGB Device							
CRT imagers	CRT and 35mm camera	Yes		>1200dpi	35mm	✓✓✓	$1,500–$3,000
Laser writers	Image written directly to film	Yes		>4000dpi	8 × 10	✓✓✓✓✓	$20,000

NOTE: Prices as of August 1995.

printer resolution and required tonal resolution to a halftone screen resolution, use the following equation:

$$lpi = \frac{dpi}{\sqrt{tonal\ levels\ excluding\ white}}$$

In the past, high quality maps and images required the use of expensive offset lithography printing, a three- or four-color process using etched metal plates. Offset lithography was only economical with very large print runs. Digital printing technologies have now made it possible to generate high-quality output of color images at relatively low cost when only a few copies are needed. The choice of digital output device depends on

- the level of detail, or resolution, required;
- whether the desired output is slides or paper copies;
- how many copies of the image are required; and
- whether the image is to be reproduced in color or black-and-white.

Digital output devices vary greatly in price, from dot matrix printers costing \$150 to large-format color electrostatic plotters at over \$50,000 (table 7.8). As a general rule, the quality and final size of the image that can be reproduced is directly related to the price of the device. Although most users will not be able to afford to buy a high-end graphics printer, commercial graphics printing companies provide the opportunity to reproduce very high quality prints and transparencies of images or maps generated digitally. Before sending images to be printed, it is well worth printing a proof strip from the image. The proof strip should vary CMYK levels, as well as brightness and contrast. High-end image processing software, such as TNT MIPS, automatically generates color and contrast proof pages from an image strip, thus ensuring that the final printed image has the correct color balance and contrast (Skrdla 1992a).

For day-to-day use, a color ink-jet or color laser printer should provide most users with acceptable hard copies of the images and maps that they generate.

8

Applications of Remote Sensing Imagery

This final chapter is designed to expose the reader to a few examples of how other researchers or agencies have made use of remote sensing information to better understand natural systems and to improve resource management. We could perhaps provide the reader, merely, with a bibliography of relevant articles in key remote sensing and photogrammetry periodicals. However, academic articles often do not provide the reader with a clear understanding of what motivated the authors' choices of imagery, the problems they had to overcome to undertake their studies, and what they would have done differently with hindsight.

We believe that the best way to learn how to use remote sensing imagery is to (1) undertake a pilot study, and (2) learn from other users. To facilitate the latter, we selected a few articles that show a range of natural resource management uses of remote sensing imagery, and then we interviewed the authors. We asked them to expand on their rationale for the methods they used, thus providing important background information to our summaries of their papers. Our discussions with both long-time and recent users of remote sensing imagery allow us to

- identify a range of key factors that first time users must address, or at least be aware of, to make effective use of remote sensing data;
- highlight pitfalls to avoid; and
- offer additional unpublished methodological advice.

Rapid Updating of Maps and GIS Data Layers

U.S. FOREST SERVICE

The U.S. Forest Service Pacific Southwest Region needed to update its 1980s 1:24,000 series of aerial orthophotomaps (geometrically corrected and mosaicked aerial photographs) and to integrate this new information into its geographic information system (GIS) for the region. Aerial photography was considered too expensive and could not generate the imagery over such a large area in a timely manner. Panchromatic 10m resolution data from the SPOT satellite were chosen because they combined rapid, wide area coverage with the capacity to generate geographically corrected 1:24,000 orthomaps.

The corporation that sells SPOT images has a product line called SPOTView, which uses USGS 7.5 minute series topographic maps and 1:24,000 digital elevation models to geometrically correct the raw panchromatic or HRV imagery. The ortho-corrected images produced are distributed as digital raster datasets for inclusion as a data layer in a GIS. They are also distributed as photographic prints in standard 7.5 minute map series format.

Using SPOTView the U.S. Forest Service was able to update its information on the size and location of clearcuts, in a format compatible with the standard 7.5 minute aerial orthophotomaps used previously, but with the added advantage of providing both standard paper maps for field use and a digital dataset to update the Pacific Southwest Region GIS.

Mapping of Habitat Suitability

AERIAL VIDEOGRAPHY IN WILDLIFE HABITAT STUDIES
(Sidle, J. G. and J. W. Ziewitz. 1990. *Wildlife Society Bulletin* 18:56–62)

Sidle and Ziewitz were interested in quantifying the nesting habitat available to piping plover (*Charadrius melodus*) and least tern (*Sterna antillarum*) along 400km of the Platte river in Nebraska, between Lexington and the confluence with the Missouri River. These birds prefer to nest on barren or sparsely vegetated sandbars along the river. The size, location, and nesting suitability of sandbars varies over time, as river flow and water level change.

Sidle and Ziewitz realized that to detect temporal changes in sandbars over a long stretch of river, with sufficient detail, field surveys were likely to be too expensive in terms of manpower and time. They decided instead to use aerial videography because it would allow information to be gathered rapidly and repeatedly over the full 400km of the survey area, at a resolution defined by them, and at a cost ($275 per flight) considerably less than that of more conventional 35–70mm aerial photography ($9,000 per flight). In addition, they selected videography because it offered the possibility of semiautomated detection and analysis of potential nesting sites through digital processing of video frames.

To conduct a videography survey, they mounted a color video camera aimed vertically through a hole cut in the baggage compartment of a Cessna 172 aircraft. The camera was attached to an NTSC video player/recorder and a small color monitor. The video signal from the camera was recorded using the highest quality recording speed, providing two hours of data on each videotape. At a ground speed of 90 knots (172km/hr), a complete survey of the 400km river section required 2.4 hours of tape. The system was powered with a 12-volt deep-cycle gel battery. The monitor was located in the cabin; the VCR and camera are operated by the pilot using a remote control. Flights were conducted in early morning or late afternoon to avoid glare from the specular reflection off the water surface. The flat topography of the area meant that shadow effects of the low sun angle were not a problem. The Cessna was flown at a constant altitude of 1,676m, with the camera zoom lens set at wide angle (10.5mm). To maintain image geometry and minimize distortion, river bends were navigated with wings-level flying by slipping the aircraft with rudder and opposite ailerons. A written log of tape-counter numbers and landmarks was used to index the videography to known locations along the river.

Using an IBM PC AT–compatible computer and a video frame grabber, 94 sample sites were extracted from the videotape. These sample scenes were classified visually into three cover types (nonvegetated sandbars, vegetated islands, and water), using the Map and Image Processing System (developed and marketed by MicroImages, Inc. in Lincoln, Nebraska). Once classified, the area of each cover type was determined. Breeding site area estimates from videography were comparable to estimates made from 35mm color photography.

Using these methods, Sidle and Ziewitz were able to examine the impacts of changing water levels associated with release of water from upstream dams and irrigation systems. Those impacts include effects on the size, location, and suitability of plover and tern nesting sites. By analyzing their aerial videography data, they were able to determine the ranges of water levels that would not jeopardize plover and tern breeding success.

MAPPING MUSKOX HABITAT IN THE CANADIAN HIGH ARCTIC WITH SPOT SATELLITE DATA

(Pearce, C. M. 1991. *Arctic* 44:49–57)

Pearce wanted to assess the area of muskox habitat on Devon Island in the Canadian high arctic. To do this, she had to detect and map small islands of productive sedge meadow isolated within a vast matrix of sparsely vegetated polar desert. Sedge meadows constitute the sole feeding patches for muskox, and their size and integrity over time is crucial to the health of the muskox population on Devon Island.

Pearce decided to use satellite imagery because aerial photography would have been very expensive for the large area to be surveyed. The small size of sedge meadows precluded use of coarse resolution AVHRR and Landsat MSS imagery. Pearce considered using Landsat TM imagery because it offers higher spectral resolution than does SPOT HRV. However, because the phenology of sedge meadows varies considerably with timing of spring and summer temperatures, she felt it was more important to synchronize image acquisition with field sampling. This consideration made SPOT HRV the better choice; its off-nadir capability provides more frequent overpasses than does Landsat TM, so there was a greater probability of obtaining cloud-free SPOT imagery during the field sampling period. By chance Pearce obtained a clear SPOT image on the exact day of intensive field sampling.

Pearce emphasizes the importance of field sampling (ground-truth data collection): unless one has a thorough understanding of the spectral characteristics of the terrain, it will be impossible to decipher the spectral information contained within the satellite imagery. Field sampling is particularly important in the arctic because vegetation patches are small and sparse—often comprising less than 50% of the landscape. Field sampling is also important because low sun angle generates shadows that alter the spectral characteristics of land-cover types. For example, ungrazed meadows appear darker than do grazed meadows, because the longer sedges cast shadows obscuring the light-colored soil that comprises much of the spectral signature of the terrain.

Using old aerial photographs and a 1:5,000 vegetation map that was fifteen years old, Pearce selected two ground sampling transects (each 6km long) within one of eight lowlands used by muskox along the north-east coast of Devon Island. Percent cover of vascular plants, lichens, mosses, litter, and bare ground were estimated using line intercept sampling along a series of randomly located 100m segments stratified within meadow, beach ridge, and rock outcrop plant communities. Above-ground biomass of vascular plants was estimated from forty 50×50cm quadrats. Information on the color of terrain features and the effect of shadows was also gathered. Pearce visited all other lowlands via helicopter to determine whether terrain characteristics were comparable to the ground-sampled sites, and to count the number of grazing muskox.

The SPOT HRV image was examined on a PCI EASI/PACE image analysis work station. A red/near-IR ratio image (band2/band3) was generated as a rough index of the amount of vegetation within the area. Using the field sampling information and the land-cover classification scheme of the published vegetation map, six evenly distributed sites were identified within a 512×512 pixel window for each of the following known types of land cover: hummocky sedge/moss meadow, frost boil sedge/moss meadow, cushion plant–moss/lichen on beach ridges, dwarf shrub/heath on granite rock outcrops, moss/herb on dolomite rock outcrops, open water (clear, silty, and shallow), and ice. These sites were used to "train/supervise" a maximum likelihood classifier, which was then used to assign the SPOT spectral data into land-cover classes. Each land-cover training site contained at least forty pixels from the largest and most homogeneous area.

The supervised classification resulted in eight land-cover classes corresponding to the five vegetated land-cover types characterized in the training sets, with the water class subdivided into three classes (ice, clear water, and shallow water/ice). Only 2.76% of the image was not assigned to a land-cover class, and this was primarily associated with disintegrating ice. The absolute and relative area of each land-cover type within lowland areas was determined from the classified image. Accuracy of the supervised land-cover classification was determined by visual comparison with the published vegetation map, and by assessing differences in areal extent from those in the literature.

Accuracy was deemed sufficient to judge this supervised classification a success in calculating the overall area of sedge meadows—the favored foraging areas of muskox—within coastal lowlands. However, spectral overlap introduced errors in distinguishing frost boil sedge/moss communities from the hummocky kind. The analysis correctly estimated the area of beach ridges, granite outcrops, and lakes and ponds.

By combining field sampling with digital analysis of SPOT HRV imagery, Pearce was able to determine that sedge meadows available to muskox along the northeast coast of Devon Island comprise only 9% ($50km^2$) of the landscape. The ability to detect, identify, and measure sedge meadows using SPOT image analysis provides the means to estimate the total area and distribution of muskox habitat throughout the vast regions of the Canadian high arctic.

AERIAL THERMAL INFRARED IMAGING OF SANDHILL CRANES ON THE PLATTE RIVER, NEBRASKA

(Sidle, J. G., H. G. Nagel, R. Clark, C. Gilbert, D. Stuart, K. Willburn, and M. Orr. 1993. *Remote Sensing of Environment* 43:333–341)

Sidle et al. were interested in determining whether the number and distribution of roosting sandhill cranes along the Platte River could be ascertained using a military infrared reconnaissance system, and whether roosting numbers are affected by changes in river characteristics associated with reservoir and water diversions in Colorado, Wyoming, and western Nebraska. During the spring migration, most of the North American population of sandhill cranes (400,000+) spends four to six weeks in March and early April on the Platte River. The cranes roost at night in river channels that are shallow and wide, with sparsely vegetated shorelines.

The Nebraska Air National Guard generated a high-resolution film map of the Platte River using an AN/AAD-5 thermal infrared sensor on board a reconnaissance F-4 phantom jet. Imagery was obtained starting at 7:15 P.M. (CST) on 28 March 1989, and 7:50 P.M. on 5 April 1989. The aircraft traveled at about 300 knots, navigating over the Platte River with the aid of on-board forward-looking radar. The thermal IR sensor is a scanning optical system that focuses IR radiation upon two detector arrays providing both a narrow 60° field of view (30° on both sides of nadir) and a wide 120° field of view. Detector arrays are sensitive to thermal IR radiation in the 3–12μm waveband and are sealed within a vacuum Dewar assembly maintained at −193°C. Sensor output is displayed on a 13cm monitor (CRT); light from which is reflected and focused onto 13cm Kodak RAR2494 film, producing a continuous record/map of the terrain directly below the aircraft. Film transport speed is controlled by a velocity-to-height ratio signal from the Aircraft Parameter Control. Latitude, longitude, and radar altitude are recorded once every 0.3m of film. At an above-ground height of 400m, the IFOV is 0.14m (60°) and 0.29m (120°). Thermal sensitivity is 1.5°K when measured against a 300°K background for objects the same size as the IFOV.

Contrast between the thermal infrared emissivity of sandhill cranes and the background (water) was sufficient to detect cranes along the river. Cranes appear to the naked eye as small gray dots on the film.

To determine the number of roosting cranes, the river from Highway 283 to Highway 34 was divided into ten segments, using bridges as dividers. Length of each segment was determined from USGS 7.5 minute series maps. A scale was determined for each segment of film strip to compensate for variance in aircraft altitude. Roosts within each bridge segment were counted. A roost was defined as a congregation of cranes separated by at least 100m from the next congregation. The area of each roost was estimated using a dot-grid overlay. The average density of cranes was estimated by counting cranes within six roosts randomly selected from three of the ten segments. A binocular microscope with eyepiece graticule was placed over the film on a light table, and the number of cranes was counted in a graticule-defined area of 5×5mm situated within each roost. The actual area censused within each roost was determined from the scale of the film within that particular segment. The average density of cranes within roosts was computed from the eighteen samples and was found to be one crane per 3.5m^2. The number of cranes for each roost site in all river segments was determined by dividing the roost area by the average crane density. River channel width was estimated for each roost by averaging channel width at three locations along the roost. Number of river channels (≥24m in width) and the distance from each crane roost to the nearest feeding and mating area of wetland meadow were measured from a set of 1984 color infrared aerial photographs (scale 1:24,000).

The thermal remote sensing survey produced an estimate of 169,391 roosting cranes on 28 March, and 285,850 cranes on 5 April. Sidle et al. noted that their estimates were lower than those of Solberg (326,995), who conducted daytime aerial survey transects on 28 March. Discrepancies on the thermal imaging estimates may have arisen for the following reasons: (1) around bends in the river, banking of the aircraft caused too much distortion for cranes to be detected; (2) image quality was also degraded by erroneous velocity-to-height ratio information; and (3) poor navigation sometimes placed the aircraft away from the river so that roosts may have been missed.

Regression analysis with number of crane roosts as the dependent variable showed distance to wetland meadows, number of islands, and number of channels to be significant factors. Channel width and braiding index (index of the number of channels) were nearly significant.

Sidle et al. were able during this exploratory study to demonstrate that thermal infrared remote sensing conducted aerially at night is capable of locating crane roosts and of censusing crane density along the Platte River. They suggest that this method would be a valuable check on the present daytime visual aerial count of feeding and loafing cranes. Furthermore, unlike diurnal surveys, thermal infrared remote sensing provides detailed information on the number, location, and size of crane roosts—variables that can be used as indices of river habitat conditions over time. Sidle et al. note that the AN/AAD-5 system is being discontinued by the U.S. Air Force and can be lent or transferred to public agencies.

Habitat Change and Habitat Loss

MONITORING FLOODS WITH AVHRR

(Barton, I. J. and J. M. Bathols. 1989. *Remote Sensing of Environment* 30:89–94)

In April 1998 heavy rains in the upper catchment of the Darling River in Queensland and northern New South Wales, Australia, resulted in considerable flooding along the entire length of the river. Barton and Bathols were interested in determining whether AVHRR imagery could be used to track the movement of the flood down the Darling River during the months of May and June. They chose AVHRR imagery because they had free access to data from the CSIRO ground receiving station that captures images during the twice-daily (daytime and nighttime) passes of the NOAA satellites. With two satellites in orbit, acquisition of two daytime images is possible. An initial day and night pair of AVHRR images was obtained over the Darling River study area on 29 May 1988. To discriminate land from water surfaces within the daytime imagery, Barton and Bathols used the normalized difference vegetation index (NDVI),

$$NDVI = \frac{near\text{-}IR - green}{near\text{-}IR + green},$$

that combines channel 1 (green) with channel 2 (near-IR) such that variations in solar illumination of the landscape are accounted for. Channel 4 (thermal IR) within the nighttime imagery was transformed to temperature. From these images the average NDVI and temperature were calculated for areas known to be upland, flooded, or lakes (deep water). (Average NDVI = [channel 2 − channel 1] / [channel 2 + channel 1].) Barton and Bathols determined that the nighttime channel 4 thermal data provided a better land-flood discrimination than did the daytime NDVI. Temperature data during the day are poor at discriminating flooded areas because both lakes and flooded areas exhibit temperatures similar to that of the surrounding land. At night, under clear sky conditions, the radiative heat loss from the land cools it down well below that of both flooded terrain and lakes.

To monitor the progress of the flood down the Darling River, three additional nighttime AVHRR images were obtained on 6, 16, and 29 June. The leading edge of the flood was determined by examination of channel 4 temperature images. Estimates of the area flooded were made by dividing the channel 4 temperature image into small areas of 64×64 pixels, selected to exclude lakes. A frequency histogram of pixel temperatures was then generated for each 64×64 pixel window. Typically, histograms showed a bimodal distribution, with the largest peak falling at the cool end of the spectrum. The other, smaller, peak was located at the warmest end of the spectrum. Based on the known average nighttime temperatures of land and deep water, Barton and Bathols concluded that the large peak represented land, the small peak was permanent (deep) water, and the area in between was flooded. From the assumption that both land and water pixels have a symmetric distribution with temperature, Barton and Bathols were able to reliably estimate the ratio of land to water in the flood pixels. Their overall estimate of the area of land flooded along the Darling River was deemed to have an accuracy of ±25%.

SATELLITE REMOTE SENSING OF TOTAL HERBACEOUS BIOMASS PRODUCTION IN THE SENEGALESE SAHEL: 1980–1984

(Tucker, C. J., C. L. VanPraet, M. J. Sharman, and G. Van Ittersum. 1985. *Remote Sensing of Environment* 17:233–249)

Tucker, of the Goddard Space Flight Center (NASA), along with VanPraet, Sharman, and Van Ittersum, of the Pastoral Ecosystems Project (FAO/UNEP), were interested in assessing forage biomass production in the Sahel region of Senegal and in determining how production varies from year to year. The Sahel is an ecological transition zone that lies between the Sahara desert and the humid savannas of sub-Saharan Africa. The Sahel zone is bounded by the 200 and 400mm isohyets. Its rainfall pattern is monomodal, with a single growing season from mid-June through September. The region's vegetation is characterized by thorny acacia and a herbaceous cover of annual grasses. Pastoralism is the major land use and source of income in the Sahel. Grasses comprise 95% of forage for cattle at the beginning of the growing season, falling to 55% at the end of the dry season as acacia browse becomes more important. Primary production depends primarily on rainfall and only secondarily on soil condition.

Primary production in the Sahel can be measured by destructive sampling in the field. Though this form of sampling provides accurate local estimates of production, the cost of field surveys and the variability in rainfall pattern over large areas makes regional estimates of production from a few local surveys untenable. Tucker et al. realized that primary production in the Sahel and elsewhere can be measured on a regional scale using satellite remote sensing techniques. The biomass of plant material can be estimated nondestructively by way of multispectral remote sensors. This is because the red and near-infrared light reflected from plants is a function of the quantity of photosynthetically active compounds present, which is in turn related to overall plant biomass. Primary production can be determined spectrally by calculating the integral of a set of red/near-IR "greenness" estimates over the growing season. Tucker et al. note that the advantage of estimating primary production using satellite remote sensing is that large areas can be measured directly and nondestructively, and that green leaf biomass estimates combine many biotic (species-specific production efficiency, etc.), and abiotic (rainfall, soil nutrient status, etc.) factors that affect primary production.

To estimate primary production in the 30,000km^2 Ferlo area of the Sahel of Senegal, Tucker et al. needed a source of frequently collected satellite data and a set of field verification samples to calibrate the remote sensing biomass estimates. Tucker et al. first looked at the availability of Landsat MSS over the region but found that there were insufficient cloud-free images for a multitemporal dataset during any single growing season. They then turned their attention to AVHRR local area coverage (LAC) data. Though LAC data have a pixel size of about 1.1km, Tucker et al. believed that this was not a constraint, as the smallest patches of primary production in the region occur in dune and interdune areas that are 1–2km wide. The major advantage of AVHRR data over Landsat MSS is its high temporal resolution, with data being available several times during every nine-day orbital cycle.

The Ferlo area of Senegal was chosen for the study because between 1980 and 1984 the Pastoral Ecosystems Project had been conducting systematic ground-based estimates of biomass. Accurate estimation of "greenness" using the Normalized Difference Vegetation Index (NDVI) requires cloud- and haze-free imagery. Between 1980 and 1984 two AVHRR sensors were in orbit on satellites NOAA-6 and NOAA-7. All AVHRR data from NOAA-6 are obtained at 7:30h local time as the satellite moves north to south. Data from NOAA-7 are obtained at 14:30h as the satellite heads south to north. The NOAA-6 imagery acquired in the early morning suffered substantially greater cloud and haze cover than did the early afternoon data from NOAA-7. As no ground receiving station existed over the study area at the time, acquisition of images was constrained by competition for on-board tape recorder space. As a result, off-nadir images were used. To control for IFOV differences associated with increasing viewing angle off-nadir (oblique), all AVHRR images were mapped interactively to a Mercator projection, using the coastline as a reference. An interactive mask based on channel 4 (NOAA-6) or channel 5 (NOAA-7) thermal IR temperatures was used to eliminate discrete patches of clouds within each image. Geocorrected multispectral images were then converted to single-layer NDVI images. When cloud cover was severe within individual scenes, artificially cloud-free images had to be created by taking the maximum NDVI value of each pixel within the three images (one nadir and two off-nadir) obtained during every nine-day orbital period that were closest to the center of the Ferlo study area. As clouds, aerosols, and water vapor tend to have negative index values, maximum NDVI value compositing (Holben 1986) is an effective way to remove most atmospheric effects—as long as each pixel is cloud-free in at least one of the composite images.

Extensive field data were gathered during and at the end of the growing (rainy) seasons from 1980 through 1983. Field data were not gathered in 1984 because of the extreme drought. An average of 72 locations were sampled each year. At each location, noted by dead-reckoning and map triangulation on a 1:200,000 vegetation map of the area (accuracy ±1–2km), six to ten $1m^2$ quadrats were selected. Within each quadrat the following data were recorded: (1) dominant and second-dominant species; (2) spectral reflectance using a hand-held, two-band radiometer; (3) percent cover; and (4) mean grass height. Each quadrat was then clipped to the ground, and the wet sample weighed. Dry weights were estimated for about half of the quadrats. (The hand-held, two-band radiometer combined a Tektronics digital radiometer with a Hewlett-Packard pocket calculator, and was designed to measure radiance in visible red, 0.650–0.700μm, and near-IR, 0.775–0.825μm, wavebands that overlap respectively with AVHRR band 1 and 2. Radiometer readings were used to convince the scientific community that spectral variability existed in relation to herbaceous biomass, and was not an artifact of sensor or atmospheric errors.)

Total annual precipitation varied substantially in the Ferlo area from 1980 to 1984, as did the total green leaf biomass estimated from destructive field sampling. The Normalized Difference Vegetation Index measures of "greenness" were integrated over time. When NDVI values were regressed against destructive sampling estimates of end-of-season biomass, there was a clear linear relationship between integrated NDVI and green leaf biomass (r^2 of 0.69). By applying the regression-derived equation to the average integrated NDVI value for the Ferlo region, Tucker et al. were able to estimate total annual primary production. Sources of uncertainty in the estimates arise from a variety of causes: image coregistration or ground sampling location errors; insufficient destructive samples; possible herbivory in areas prior to end-of-season destructive sampling; elevation of NDVI estimates by haze and subpixel clouds; and interobserver variance in destructive sampling.

Marked variations in the spatial distribution of green biomass were observed both within and across years. In the zone of 100–200mm mean annual precipitation, total primary production for the year can occur within a very short period (2–3 weeks). The high temporal resolution of the AVHRR sensor is more crucial, therefore, than spatial resolution for detecting such brief pulses of productivity. Correlation of integrated NDVI and end-of-season biomass works well in the 250–500mm annual precipitation zone. Below 250mm the technique shows a lack of sensitivity. To improve the sensitivity of NDVI/biomass relations to transient primary production in more arid regions of the Sahel, Tucker et al. calculated the maximum NDVI for each pixel over the growing season, and regressed these data against destructive sampling end-of-season biomass estimates. Results, as hoped, provided quantitative information on green leaf biomass below the 250mm precipitation isohyet (r^2 of 0.64).

Using these two techniques, the authors were able to demonstrate that AVHRR data can be used to map and monitor primary production over large areas of the Senegalese Sahel, and potentially over the entire Sahel zone.

FEASIBILITY OF USING SATELLITES FOR DETECTION OF KINETICS OF SMALL PHYTOPLANKTON BLOOMS IN ESTUARIES: TIDAL AND MIGRATIONAL EFFECTS

(Tyler, M. A. and R. P. Stumpf. 1989. *Remote Sensing of Environment* 27:233–250)

Tyler (Altalo) and Stumpf were interested in demonstrating that satellite image analysis could produce reliable and accurate measurements of chlorophyll distribution in a turbid estuary, with the goal of developing a tool to monitor algal bloom migration, origins, and dissipation. The authors chose the Chesapeake Bay and its major tributaries as a study site, because (1) algal blooms are a common feature of this nutrient-rich and turbid estuary, (2) many of the factors controlling bloom formation have been determined, and (3) in situ data had already been collected to examine fine-scale bloom dynamics of estuarine dinoflagellates.

The dinoflagellate *Heterocapsa triquetra* forms algal blooms in late fall and early spring. Its motile stage constitutes a major food web dietary item, contributing to the high productivity of the estuary in late winter. Bloom distributions reflect water flow and salinity patterns within the estuary.

Satellite imagery was obtained from the experimental Coastal Zone Color Scanner (CZCS) flown on the NIMBUS-7 satellite, and from the

AVHRR on-board the NOAA polar-orbiting weather satellites. The CZCS was operational from late 1978 to mid 1986; it has a spatial resolution of 0.825km and a revisit interval of 2–3 days at midlatitudes. Imagery from AVHRR has a 1.1km pixel size, with NOAA-6 providing morning (0800h local time) and NOAA-7 providing afternoon (1400h local time) imagery. The AVHRR imagery was chosen for its twice-daily revisitation capability. The CZCS imagery was chosen because sensor wavebands and gain settings were designed specifically to image coastal waterbodies.

Water color was determined as the ratio of the variation in water column reflectance,

$$C_{ji} = \frac{\Delta R_j}{\Delta R_i},$$

where ΔR is the deviation from the clear water reflectance. For visible red (i) and near-IR (j) radiation,

$$\frac{\Delta R_j}{\Delta R_i} = \frac{R_j}{R_i}.$$

AVHRR bands 1 (0.58–0.68μm) and 2 (0.725–1.1μm), and CZCS bands 4 (0.66–0.68μm) and 5 (0.70–0.80μm) were used for the analysis. Band 4 of CZCS was selected because it corresponds to the radiation absorption peak of chlorophyll-a. Band 1 of AVHRR is broader and is influenced by other algal pigments, chlorophyll-b and -c, though chlorophyll-a absorption should still dominate band 1 brightness levels (Stumpf and Tyler 1988).

Vertical profiles of algae, temperature, salinity, chlorophyll-a, suspended sediments, and nutrients had been collected at twenty stations in the bay and lower reaches of the Potomac River from 12–15 April 1982. To make use of these in situ data, the authors obtained two AVHRR images on 14 April, and one CZCS image on 15 April 1982. Images were resampled using a nearest-neighbor technique and mapped to a Mercator projection. An atmospheric haze correction was undertaken by subtracting from all pixels the minimum radiance observed within deep, clear, shelf waters. The authors note, however, that this method assumes a homogeneous atmosphere, and can generate errors associated with the increasing atmospheric path length of pixels with increasing deviation from nadir. This is because the radiance of deep-water pixels is likely to increase with increasing atmospheric path length, which is a function of the pixel's deviation from nadir. Given this, a subtraction atmospheric correction using the minimum radiance from an off-nadir deep-water pixel will tend to correct off-nadir pixels but will overcorrect near-nadir pixels. The converse is true for subtraction corrections using near-nadir pixels.

The authors found that the color index (C_{ji}) did not vary with reflectance (i.e., changes in water color owing to turbidity and sediment load) but only with chlorophyll concentration. Furthermore, the distribution of very high color index (C_{ji}) values located within the Potomac River corresponded closely to a bloom identified by shipboard observations. The color index is, however, sensitive to atmospheric conditions, showing increased values with increasing haze and clouds. The high gain of CZCS sensors resulted in the visible red channel (band 4) becoming saturated over regions of high turbidity (high reflectance). (Saturation of a sensor occurs when the brightness of a pixel exceeds the gain of the sensor. The brightness value of such pixels is set to the maximum of the sensor, and thus variance in brightness that exists among saturated pixels is lost.) As a result, the AVHRR sensor proved more useful than the CZCS in characterizing water color in high turbidity areas.

In situ estimates of chlorophyll concentration were regressed against the color index data computed from the AVHRR and CZCS imagery, with correlations of 0.96 and 0.95 respectively. The resulting functions were then used to transform the color index AVHRR and CZCS imagery to chlorophyll distribution maps of Chesapeake Bay.

Examination of the multitemporal AVHRR dataset showed diurnal changes in chlorophyll concentration within the area occupied by the bloom. Values of $<50\mu g/L$ were estimated from a 0810 EST NOAA-6 AVHRR image on 14 April 1982. That value changed to $>100\mu g/L$ for a 1410 EST NOAA-7 AVHRR image on the same date. Though some of the disparity may have been a result of sensor differences, the two satellites gave identical water column reflectances over the rest of the bay. The dinoflagellate responsible for the bloom is phototactic and exhibits strong vertical migrations. This fact was demonstrated both by in situ observations of the vertical distribution of cell concentrations within the water column (90% of cells within the surface meter at midday, compared to half that in the early morning) and by the color index changes detected between the 0810 and 1410 EST images. Examination of the 1410 EST

AVHRR image obtained at low tide, and the CZCS image obtained at 1530 on 15 April at high tide, showed movement of the bloom upstream with the tide, and a cross-channel expansion of the bloom at high tide.

By combining knowledge of dinoflagellate blooms and satellite sensor characteristics with in situ data collection in Chesapeake Bay, Tyler and Stumpf were able to demonstrate the effectiveness of AVHRR and CZCS imagery for studying algal blooms in turbid estuarine waters. The authors showed that AVHRR and CZCS data can be used to generate a water color index that is highly correlated to chlorophyll concentrations in the surface meters of the bay, and that the high temporal resolution of the AVHRR satellite system makes it possible to detect and map the distribution of algal blooms and to monitor their diurnal and daily movements within the estuary.

RGB-NDVI COLOR COMPOSITES FOR VISUALIZING FOREST CHANGE DYNAMICS

(Sader, S. A. and J. C. Winne. 1992. *International Journal of Remote Sensing* 13(16): 3055–3068)

Softwood forests are a major economic resource in Maine. Legislators and the private forest industrial sector are forced, however, to make resource management decisions and policies based on insufficient or outdated forest resource information. Sader and Winne were interested, therefore, in determining the feasibility of using Landsat MSS (data more than two years old are available from EROS for $200) and Landsat TM image analysis to help the Maine Forest Service conduct timely and cost-effective statewide forest surveys and forest change monitoring. Specifically, Sader and Winne wanted to demonstrate a new and simple technique to display and quantify forest change at five- to ten-year intervals that correspond with the U.S. Forest Service survey cycle.

Satellite imagery was obtained for three dates: September 1978 and June 1981 Landsat MSS and June 1987 Landsat TM. The TM data were registered geometrically to a UTM projection, and the corner points of the study area were located (all in the Great Pond, Maine, 1:24,000 quadrangle). The study area was chosen because it represents a standard map reference format, and includes private industrial forest land where forest harvesting and regeneration can be examined over the time frame of a forest survey cycle. The visible red and near-IR wavebands were extracted from all the imagery. The visible band was corrected for haze by locating the minimum data value from deep clear water and subtracting this from each pixel in the red waveband. No correction was needed in the near-IR, as the minimum data value was already zero. The MSS images were then destriped to correct sensor gain errors. Matching points unambiguously identifiable in the geometrically corrected TM image and in both MSS images were located and used to geometrically correct the MSS images that were also resampled to match the TM image resolution of 30m.

The Normalized Difference Vegetation Index (NDVI) was computed for each dataset. A linear contrast stretch that cropped the tails of the distribution (only data two standard deviations from the mean were included in the stretch) was then performed on each NDVI image to create an 8 bit image (0–255 brightness levels). The authors found that this saturation contrast stretch proved effective in distinguishing vegetated from nonvegetated areas in Maine. Haze correction, ratioing, and scaling helped reduce the differences in illumination and in sensor gain and offset likely to exist among the images. The index was computed to reduce the volume of data, but to retain a quasi-continuous variable that has been shown to represent green leaf biomass. Major decreases or increases in NDVI are likely to be associated with tree harvesting or regeneration.

The authors rationalized that they could use the principles of additive color theory to combine the three NDVI images from different dates into a composite color image that would show changes in forest color. (According to these principles, equal brightness of red and green with no blue produce the color yellow; absence of red, green, and blue produce black; and high values of red, green, and blue generate white.) To visualize and analyze forest change across the three-date time series, the authors conceived of the novel idea of assigning the three NDVI images to the red, green, and blue color guns in a standard RGB display to generate a color composite image. They assigned NDVI image 1978 to blue, 1981 to red, and 1987 to green.

With an understanding of additive color theory, and knowing which color gun was assigned to which date image, the colors in the composite image and their meaning could be interpreted. For example, a high NDVI value in the 1978 image (blue), and low values in 1981 (red) and 1987 (green), would result in those pixels appearing blue in the composite RGB image. Roads that were built after 1978 appear blue in the composite image because they were vegetated in 1978 (high NDVI; blue) and nonvegetated (low NDVI) in the 1981 and 1987 images. Some colors represent more than one direction of change. For example, areas logged between 1978 and 1981 followed by regrowth/afforestation after 1981 would result in high NDVI in 1978 (blue), low NDVI in 1981 (red), and high NDVI in 1987 (green), and would appear as blocks of cyan (blue + green) pixels in the composite image.

Using a modified parallelepiped routine in the remote sensing analysis software ERDAS (RGBCLUS), the three-channel composite image was clustered into 27 color classes, which were clumped interactively into 9 subjective change classes based on the users' knowledge of the study area. The color scheme for the final 9-class image was chosen to retain the additive color logic, thus facilitating visual interpretation. To remove isolated pixels that gave the image a peppered appearance, the classified image was smoothed, with insignificant loss of information, using a 5×5 majority (mode) filter.

Sader and Winne were able to use the smoothed-change detection image to visualize and quantify forest dynamics in the study area from 1978 to 1987. The success of this exploratory study suggests that this method could be useful to forest managers in the public and private sector to monitor forest harvesting, regeneration, and pest damage on a five-year U.S. Forest Service inventory cycle.

Sader and Winne suggest that satellite images should be obtained in the same season, in order to reduce NDVI differences associated with phenological variability. They also suggest that for forest surveys data should be obtained before herbaceous growth and agricultural fields attain high levels of green biomass. The authors note also that combining the RGB-NDVI image with additional data layers—such as land cover, soils, and ownership—within a GIS would improve interpretation and quantification of forest change.

THEMATIC MAPPER DETECTION OF CHANGES IN THE LEAF AREA OF CLOSED CANOPY PINE PLANTATIONS IN CENTRAL MASSACHUSETTS

(Herwitz, S. R., D. L. Peterson, and J. R. Eastman. 1990. *Remote Sensing of Environment* 29:129–140)

Leaf area index (LAI) is a standard ecological expression of the leaf area of a plant community per unit area of terrain. It is a reflection

Table 8.1
Interpretation of a Three-Date RGB-NDVI Composite Image

		NDVI Value				
		1978	1981	1987		
Class	Additive Color	Blue	Red	Green	Green Biomass Change	% Area
0	Black	L	L	L	Water	4.8
1	Black	L	L	L	Low biomass NO CHANGE	6.5
2	Blue	H	L	L	↓after 1978	1.9
3	Red	L	H	L	↑to 1981; ↓ after 1981	3.3
4	Green	L	L	H	↑to 1987	2.7
5	Magenta	H	H	L	↓ after 1981	4.4
6	Cyan	H	L	H	↓after 1978; ↑ after 1981	2.6
7	Yellow	L	H	H	↑after 1978	6.8
8	Dark Gray	M	M	M	Softwoods NO CHANGE	22.5
9	Light Gray	H	H	H	Hardwood NO CHANGE	44.5

NOTE: L = low, M = moderate, H = high NDVI values.

of the photosynthetic and primary production capacity of that plant community. This index has been used to monitor community productivity and health over time in response to natural and anthropogenic perturbations. Leaf area index is usually estimated using log-transformed regression equations that relate leaf biomass or area (determined by destructive sampling of trees) to trunk diameter at breast height (DBH). Herwitz et al. noted that foresters managing evergreen plantations for timber or paper production would benefit greatly from the ability to monitor changes in LAI throughout the large areas covered by their plantations. Changes in LAI may reflect, for example, leaf area and productivity reductions caused by diseases, pests, pollution, or water stress.

From previous studies, Herwitz et al. knew that there was a significant correlation between Landsat TM band 4/3 ratio data and LAI. In conifer forests, which are a key source of industrial raw materials in the United States, this relationship was often confounded. Incomplete closure of the dark conifer canopy meant that the bright understory vegetation dominated the spectral reflectance of the terrain. Herwitz et al. hoped to determine whether Landsat TM 4/3 ratio data could be used to detect, quickly and nondestructively, local-scale changes in LAI known to have occurred in a pine plantation in central Massachusetts. The two pine plantations chosen for study were on level or gently sloping ground (to reduce topographic shadow effects) and consisted of mixed-species stands of red pine (*Pinus resinosa*) and white pine (*Pinus strobus*) ranging in size from 1–18ha.

Field work was undertaken in both plantations in 1987. Regression equations relating leaf biomass to DBH were estimated for the two pine species in the plantations by felling and defoliating a sample of trees of each species, with DBH ranging from 14–45cm. All fresh leaf mass was weighed. A subsample of fresh leaves was oven-dried and weighed. Leaf biomass for all sampled trees was then estimated from the known ratio of dry/fresh leaf mass, and regressed against DBH. The best-fit equation derived for each species was of the form

$$\ln(y) = A + Bln(x),$$

where y is the leaf biomass in kg, and x is DBH in cm. Surface area of fresh leaves or needles was determined (n=250) using a LI-3000 Leaf Area Meter. Measured leaves were then dried, and dry weight to surface area ratios were determined. These ratios were then used to convert leaf biomass totals to total leaf surface area. Standing LAI for the plantations was determined by summing the total leaf area of trees in sample plots ($300m^2$ and $6{,}750m^2$) and dividing by the area of the plot.

In September 1983 one of the two plantations was thinned by selective felling of intermediate-size trees. A total of 5,573 trees of known DBH were removed from the 18ha plantation. Herwitz et al. determined, retrospectively, the total leaf area removed by applying their 1987 species-specific allometric equation to the DBH recorded for each thinned tree. By dividing the total leaf area by the size of the plantation, the authors were able to estimate the overall LAI reduction associated with thinning: 28% from 4.52 to 3.24 m^2 per m^2.

Landsat TM images were acquired from a Landsat-4 overpass on 10 September 1982 and a Landsat-5 overpass on 16 September 1987. The first image was chosen because it corresponded to the time of year that the plantation was selectively thinned. Furthermore, the image had already been obtained by a colleague of Herwitz, and atmospheric conditions (low humidity, good visibility) and sun angle at that time of year were good. (Before the Land Remote Sensing Commercialization Act of 1984 became law, Landsat imagery was acquired by the USGS and distributed from the EROS data center in Sioux Falls, South Dakota; at that time, data could be exchanged freely for purposes of scientific research.) A search of the EOSAT database showed that, by chance, a Landsat TM image with very little cloud cover had been archived five years and six days after the 1982 pre-thinning image. Though Herwitz et al. had not set out explicitly to obtain anniversary images, the fact that both scenes were obtained at approximately the same time of day and time of year reduces the differences in spectral reflection of the terrain associated with differences in solar zenith angle, thus facilitating multitemporal comparisons.

The 1982 and 1987 images were coregistered, using control points unambiguously identifiable within both images. Pixels that represent each plantation were extracted by interactively delineating a mask that bounded each plantation stand. For each stand the digital number of each pixel in Landsat TM bands 3 (red) and 4 (near-IR) was converted to radiance, based on the gain and offset characteristics of the band 3 and 4 sensors on the respective satellites. Deep clear water does not reflect radiation in the near-IR wavelengths of band 4, thus any reflectance is assumed to be caused by atmospheric scattering. Atmospheric impacts on radiance from the pine stands were therefore removed from every pixel in band 4 by subtracting from each the mean band 4 radiance observed from a sample of 1,050 pixels in the deepest section of the Wachusett Reservoir, from all band 4 pixels. Band 3 atmospheric correction was undertaken using the atmospheric transmittance/radiance computer program (LOWTRAN 6) of the Air Force Geophysics Laboratory. This program requires estimates of visibility, air temperature, solar azimuth, land elevation, and day of the year. Visibility and air temperature were obtained from the Worcester Airport Meteorological Station. To account for the slight differences in solar zenith angle between the two images, band 4 and band 3 corrected radiances were divided by the cosine of the effective incidence angle of incoming radiation.

Mean corrected radiance in band 3 and 4 was calculated for the thinned plantation in both the 1982 and 1987 images. These data were used by Herwitz et al. to establish whether the differences in band 4/3 ratios mirrored the known reduction in LAI associated with plantation thinning. Herwitz et al. were concerned that the small and irregular shape of the unthinned plantation stands would lead to inclusion of a large proportion of mixed (edge) pixels in the average corrected radiance estimates.

To resolve this, the centrally located pixel within each stand was identified interactively by using the image processing system (IDIMS). A set of nine homogeneous pixels was then selected, using a polygon growing algorithm by which neighboring pixels were included in the central pixel set if they satisfied a minimum sum of differences criterion across TM bands 3, 4, and 5.

No significant correlation could be found between the mean LAI estimated for each of the unthinned stands and the corresponding mean TM band 4/3 ratio. The authors also computed an average NDVI for each of the thinned stands, but these were not significantly correlated with LAI. Absence of a significant relationship between the TM band 4/3 ratio and LAI does not necessarily mean that no relationship exists; a correlation could be masked by errors introduced when calculating either variable. Comparison of the 1982 and 1987 TM data for the thinned plantation showed a greater than 20% reduction in the band 4/3 ratio between the two dates. Herwitz et al. note that this matches very well the 28% reduction in LAI calculated using allometric equations. Contrary to the equilibrium LAI condition that should be expected in a pine plantation planted more than forty years earlier, the unthinned plantation stands showed a reduction in the TM band 4/3 ratio over time. This reduction was comparable to that of the thinned plantation. Given this, Herwitz et al. were unable to determine whether the allometric equation estimates of LAI or the TM band 4/3 ratio estimates were more reliable sources of data to monitor local changes in leaf biomass within pine plantations.

The authors thought a great deal about the factors that could affect spectral reflectance from pine plantations, and they took care to minimize errors in band 4/3 ratio values that may have arisen as a result of sun angle, topographic shadowing, atmospheric scattering, edge effects, and coregistration. Nevertheless, their results show that detecting rather subtle changes in the landscape is fraught with difficulties. Even with considerable information based on intensive field surveys and local meteorological records, the changes in TM band 4/3 ratios observed over time in both thinned and unthinned plantations could not be explained.

With hindsight, Herwitz noted that larger plantations with fewer edge pixels, along with more field data on leaf fall and canopy closure over time, might have helped demonstrate a correlation between TM band 4/3 ratios and LAI. With such added information, the authors might have been able to explain observed changes in band 4/3 ratios.

ESTIMATION BY REMOTE SENSING OF DEFORESTATION IN CENTRAL RONDÔNIA, BRAZIL

(Stone, T. A., I. F. Brown, and G. M. Woodwell. 1991. *Forest Ecology and Management* 38(3–4): 291–304)

The population of the state of Rondônia, on the western margin of the Brazilian Amazon, has soared in recent years—from an estimated 57,000 in 1950 to more than 700,000 in 1985. That increase owes to relatively fertile soils, coupled with the displacement of large numbers of sharecroppers from southern and central Brazil, caused by the mechanization of agriculture. As a result, Rondônia has experienced some of the fastest rates of tropical forest clearing in the world. Stone et al. were interested in using remote sensing imagery from the Landsat and NOAA satellites to estimate the rate, extent, and distribution of forest clearing in central Rondônia in the 1980s, and to estimate changes in road construction in the area between 1973 and 1986.

Rondônia (243,000 km^2) has a tropical seasonal climate, with a long rainy season (October–May) and a pronounced dry season during which time cloud-free satellite data can be acquired. Stone et al. noted that the enormous size of the region made it prohibitively expensive to use Landsat MSS or TM imagery as the sole source of remote sensing data with which to estimate forest clearing. They decided therefore on a hybrid strategy that would use moderate resolution Landsat imagery for local estimates of forest clearing and lower resolution NOAA AVHRR for the regional estimates.

Cloud- and haze-free Landsat-2 MSS (1980) and Landsat-5 TM (1986) images were obtained from the Brazilian Space Agency, the Instituto National Pesquisas Espaciais (INPE). This agency has a Landsat data receiving station and thus had a good catalog of minimal cloud cover imagery for the 1980s. In addition, INPE has fewer restrictions on the use of its imagery than does the U.S. distributor, EOSAT. Although both images cover an area of 185×185km, the orbital paths of the two satellites differ, thus the area overlap for both images was only 166×179km. Lastly a 1988 AVHRR image covering the whole of Rondônia was obtained with <1% cloud cover. Acquisition of suitable cloud-free AVHRR imagery was somewhat difficult at the time because it predated the automated database and microfiche data browsing system set up by NOAA in October 1990. (The Satellite Data Services Division of NOAA has two systems for users to locate useful archived AVHRR imagery. One is OSCAR, an on-line catalog of archived imagery accessible directly by modem [1-800-528-2514; set modem to Full Duplex, VT100 or ANSI emulation, No parity, 8 data bits, 1 stop bit, baud up to 9600], or via the Internet [telnet 140.90.110.5]. A second source is ILABS, an image library on optical disk that allows users to browse for useful images in the archive.)

Stone et al. used the Landsat imagery to measure two types of land use change. Changes in the total length of roads were measured by visual inspection of photographic prints of the imagery (1986 TM band 3 at 1:250,000; 1980 and 1973 MSS band 5 at 1:500,000). Visible red bands were chosen because in this portion of the spectrum healthy vegetation absorbs light and appears dark in the image, whereas roads and clearings (soils) reflect strongly and appear as light areas in the image. The very high contrast may allow even subpixel-wide roads to be detected within the imagery. But narrow roads shaded by the canopy of roadside trees would not be detected; thus, as Stone et al. noted, their estimates of road length must be considered minima of the actual road length. Stone et al. were interested in road clearing because this is often the precursor to immigration and more extensive deforestation.

Changes in the area of deforestation were determined by digital analysis of the 1986 Landsat TM and 1988 AVHRR imagery. The Landsat TM image was classified into four general land-cover classes (forest, cleared, savanna, and water), using a supervised maximum likelihood classifier within the ERDAS image-processing system. Spatially distributed training sites of several hundred pixels each that represented known areas of forest, clearing, savanna, and water were selected within the imagery, and used to generate statistics with which to classify the entire image. Results from the supervised land-cover classification of the 1986 Landsat TM image were compared to the land-cover estimates that Stone et al. generated previously by using a 1980 Landsat MSS image. The authors were interested primarily in comparing changes in the total area of forest clearing between 1980 and 1986. Though resampling the MSS and TM images to the same spatial resolution and coregistering them to detect pixel-by-pixel changes in land cover would have eliminated errors associated with different image geometry, the authors had problems reading the MSS data from tape—thus preventing direct digital comparison of the two images. The magnetic tapes used to store satellite imagery are guaranteed for only ten to fifteen years. As a result, many Landsat MSS images in the EROS and INPE archives acquired before 1980 are unreadable and in the absence of duplicates are thus effectively lost.

The AVHRR data were registered geometrically to a 1:1,000,000 map of Rondônia, with a resulting accuracy of three pixels (3.3km). If the imagery had not been corrected geometrically, off-nadir distortion of pixels at the edges of an image would have resulted in inaccurate area estimates of land cover. A mask was then used to extract the portion of the geometrically correct image representing the state of Rondônia. That extracted image was then classified into 50 spectral classes, using bands 1, 2, and 3 and an ERDAS unsupervised clustering routine. Spectral classes were relabeled to 2 major and 9 minor clearing classes, based on the authors' field experience and on available maps. The authors repeated the unsupervised classification, generating 75 spectral classes that were relabeled to 17 clearing classes. Area of clearing was then estimated from both the 11-class and the 17-class images.

Using these methods, the authors were able to demonstrate that the length of roads within a region of Rondônia near Highway BR-364 increased from 110km in 1973, to 1,410km in 1980, and 4,660km in 1986—an average annual increase of 320%. The authors noted that, though roads are not directly correlated with total area of land clearing, their analysis of Landsat imagery does indicate an astonishing increase in accessibility of once-isolated forest to immigrants over a thirteen-year period. Results of the classified Landsat image analysis showed a decrease in forest from 94% in 1973 to 82% in 1986, with the area of clearing increasing from 230km^2 in 1980 to 3,390km^2 in 1986. The spectral characteristics of natural savanna and some types of cleared land led to some classification confusion, such that natural savanna was most likely underestimated by 1–2%, and the area of clearing was thus slightly overestimated. Estimates of forest clearing for the state of Rondônia based on two different unsupervised classifications of the 1988 AVHRR data were 37,200km^2 and 37,900km^2.

Protected Area Design and Management

GAP ANALYSIS: A GEOGRAPHIC APPROACH TO PROTECTION OF BIOLOGICAL DIVERSITY

(Scott, J. M., F. Davis, B. Csuti, R. Noss, B. Butterfield, C. Groves, H. Anderson, S. Caicco, F. D'Erchia, T. C. Edwards, Jr., J. Ulliman, and R. G. Wright. 1993. *Wildlife Monographs* no. 123)

In 1982 Scott and Kepler started an intensive field study to map the distribution of threatened and endangered bird populations in Hawaii. Individual range maps were compiled and then combined to create species-richness maps for the island. When these species-richness maps were overlain with a map of parks and reserves, the authors discovered that less than 10% of the ranges of endangered forest birds were protected. Furthermore, of the forest areas with three or more endangered bird species, none were in existing parks or reserves. Scott realized that areas selected for establishment of parks and reserves were chosen not primarily to protect areas of known species richness but as a result of historical accident, absence of conflict with human populations, and low value of the land for commercial exploitation.

Crumpacker et al. (1988) found that 25% of the potential natural vegetation in the United States was under-represented on federal and Indian lands (which were assumed to provide protection). Growing concern about loss of biodiversity in Australia and a desire to establish an ecologically rational system of reserves encouraged the Australian agency in charge of national parks and wildlife to devise a quick and cost-effective method for assessing the conservation value of large geographic areas (Specht 1975; Bolton and Specht 1983; Pressey and Nicholls 1991). Borrowing from the success of Australians, and with the realization that no broad-scale assessment of the level of protection given to actual vegetation types and areas of high species richness existed for the continental United States, the U.S. Fish and Wildlife Service decided to fund a suite of studies in the lower forty-eight states to locate regions of species richness, and to identify gaps in protection of biological diversity.

Gap analysis, as conceptualized by Scott and colleagues,

> provides a quick overview of the distribution and conservation status of several components of biodiversity. It seeks to identify gaps (i.e., vegetation types and species that are not represented in the network of

> biodiversity management areas) that may be filled through establishment of new reserves or changes in land management practices. Gap analysis uses the distribution of actual vegetation types (mapped from satellite imagery) and vertebrate and butterfly species (plus other taxa, if data are available) as indicators of, or surrogates for, biodiversity. Digital map overlays in a GIS are used to identify individual species, species-rich areas, and vegetation types that are unrepresented or under-represented in existing biodiversity management areas. Not a substitute for a detailed biological inventory, gap analysis organizes existing survey information to identify areas of high biodiversity before they are further degraded. It functions as a preliminary step to the more detailed studies needed to establish actual boundaries for potential biodiversity management areas.

Gap analysis assumes that landscape diversity is likely to provide high niche diversity and thus result in high species richness. Plant community composition of a given area reflects many physical factors, such as climate, soil, elevation, and aspect, and serves as habitat for the animal community. Given this, gap analysis uses vegetation as a surrogate for ecosystems and is the foundation for assessment of biodiversity distribution.

Classification schemes As vegetation comprises the primary data layer in a gap analysis geographic information system, choice of vegetation classification scheme is exceedingly important, particularly as it must allow comparisons among all forty-eight participating states. Scott et al. determined that the vegetation classes used to map vegetation within the U.S. gap analysis project must

- be discriminable in satellite remote sensing imagery and identifiable in large- and medium-scale aerial photographs;
- correspond to, or at least be compatible with, recognized vertebrate habitat classification systems (Wildlife Habitat Relations);
- describe seral as well as climax vegetation; and
- (should) be compatible with those of adjacent states, to allow for regional and national analyses.

At present a national hierarchical classification system describing vegetation cover types, compatible at the series level with existing regional and national classifications, is being developed by the Fish and Wildlife Service, in cooperation with The Nature Conservancy. This classification will permit standardization of vegetation class names in adjacent states.

Gap analysis requires that numerous distribution maps for vegetation communities and animal species be combined and analyzed. Overlaying and analyzing the maps generated during a gap analysis would be difficult and prohibitively time consuming without the use of geographic information systems (GIS) that excel at storing, displaying, and analyzing spatial datasets. Any GIS used in gap analysis must be able to integrate both raster and vector data formats. Satellite images used to generate vegetation distribution maps are all in raster (cellular) format, whereas animal distribution data and digital topographic and soil maps are stored, typically, in vector format. For gap analysis, vector format data are preferred for representing boundaries (e.g., political, land ownership, ecoregions), for storing point data (e.g., animal sighting locations), for delineating networks (e.g., roads, rivers, corridors), and for mapping generalized land use or habitat patches (e.g., agricultural regions, ecosystem blocks, watersheds). Raster format data are preferred for storing raw and classified imagery, digital elevation data, and other continuous surfaces.

Gap analysis relies heavily on compiling existing maps, such as U.S. Forest Service timber survey maps, Bureau of Land Management maps, U.S. Fish and Wildlife Service national wetlands inventory maps, Soil Conservation Service soil type maps, U.S. Geological Survey land use and land-cover maps, U.S. Environmental Protection Agency ecoregion maps, and maps from state and local agencies—particularly from state natural heritage programs.

Before the gap analysis program began, most states did not have recent statewide vegetation maps with complete and consistent descriptions of actual vegetation types. To generate statewide vegetation maps quickly and cost-effectively, Scott et al. opted for Landsat TM image analysis. Landsat TM imagery has higher signal-to-noise ratio, radiometric resolution, spatial resolution, spectral resolution, and cartographic accuracy than does Landsat MSS imagery. But SPOT HRV imagery has even higher cartographic quality and spatial resolution than does TM data, and is acquired in late morning, thus reducing sun-angle shadow effects. However, SPOT HRV data have lower spectral resolution (particularly in the IR bands important for vegetation surveys) and are considerably more expensive for large area surveys than Landsat TM. Though the higher spatial resolution of TM relative to MSS may be important in mapping small features such as wetlands or riparian vegetation, it can actually result in lower classification accuracy by disaggregating spectrally heterogeneous landscapes (e.g., wooded suburban areas are classified erroneously into two classes—urban and forest).

Vegetation mapping Building on the experience of the two pilot projects that used gap analysis (in Idaho and Oregon), Scott et al. devised a standardized approach to vegetation mapping. Landsat TM satellite imagery acquired for gap analysis were resampled to an Albers equal-area projection with a 100m resolution. Images were classified using visual photointerpretation of three-band (4=red, 5=green, and 3=blue) false-color IR images (e.g., California), or by using unsupervised classification methods (Utah, Arizona, New England). Resulting spectral polygons were digitized (on-screen or from photographic imagery) and assigned primary and secondary vegetation cover-type attributes. The authors found that classification accuracy could often be improved by incorporating digital elevation data that can account for illumination differences associated with landscape relief and aspect, and by comparison with NASA high altitude aerial photography or low altitude aerial videography. The vegetation data layer was stored as a single, continuous statewide coverage. Ancillary vector data layers were stored as digital files corresponding to the boundaries of 1:250,000 USGS topographic map quadrangles.

Animal distributions Range maps of species found in natural history texts and field guides tend to be of extremely small scale and highly generalized; they do not exclude inappropriate habitat found within the overall distribution of the species. Range maps are compiled from point-sighting information (dot distributions), which are then extended to the boundaries of major biomes. Dot distribution maps approximate the true range of a species when the observations are numerous and recent. Historical sightings may no longer be valid if human activity has altered or transformed preferred habitat patches. State natural heritage programs have begun to develop vertebrate characterization abstracts (VCAs) that contain state-specific information on the ecology and distribution of plant and animal species.

Gap analysis also uses knowledge of an animal's habitat preferences (DeGraaf and Rudis 1986) to predict species presence or absence based on the distribution of preferred vegetation. Using habitat (vegetation) as an indicator of the probable distribution of species allows inappropriate habitat to be excluded within a given region and unexplored areas to be included as predicted range. Depending on the habitat specificity of species, vegetation-derived distributions can constitute refinement of historical range maps. Gap analysis groups structurally and floristically similar vegetation types into broader categories more meaningful to the association of species and habitat types. For example, in California 375 natural plant communities are recognized, and these cross reference to 53 wildlife habitat types; in Idaho 113 vegetation cover types were generalized into 33 broader habitat types.

Scott et al. warn, however, that several factors complicate the use of vegetation to map species distributions. Birds respond more to vegetation structure (e.g., canopy closure, height differentials, tree architecture) than to plant species composition; thus vegetation classes must attempt to include these characteristics. Animals vary greatly in the breadth of their habitat specificity. Generalists such as the coyote (*Canis latrans*) and raccoon (*Procyon lotor*) thrive within a wide range of land-cover types, whereas the specialist black-tailed prairie dog (*Cyonomys ludovicianus*) is restricted to semiarid grasslands. Furthermore, the habitat requirements of some species vary in different parts of their range.

To derive species distribution maps within gap analysis projects, four sets of information are combined within a GIS: (1) a digital map of vegetation cover types; (2) a digital map of the region divided into geographic units such as counties; (3) a database indicating the presence or absence of species in each geographic unit; and (4) a database predicting the presence or absence of animal species in each vegetation or habitat type. In the simplest case the vegetation map is combined with the geographic map to create polygons with combined vegetation and county attributes. Using the species distribution databases, one can assign to each polygon within the combined vegetation/county data layer a presence (1) or absence (0) attribute for each species in the distribution database. The GIS can then be used to display single or multiple species distribution maps or to generate species-richness maps. Scott et al. note that the coarse-scale (state or ecoregion) minimum mapping unit (spatial resolution) of the gap analysis GIS is much larger than are the often micro-scale habitat preference descriptions for individual species, and thus it can introduce generalization errors in the final distribution map.

Wetlands and riparian areas often constitute only a small proportion of the landscape but are disproportionately important as wildlife habitat. This poses certain difficulties, as these critical habitats are often very small relative to the regional mapping scales necessary for gap analysis. To resolve this problem, Scott et al. first assigned wetland-associated species to any polygon that contained wetland or riparian cover types. Though this included wetland species that would not otherwise have been mapped, the predicted distributions of these species were highly generalized, erroneously covering broad expanses of uplands. A more successful approach in Idaho mapped statewide datasets of streams and lakes onto a 1:100,000 hydrographic digital line graph (DLG). The entire statewide dataset was too large to manipulate even with a high-speed computer, so the category of smallest streams was eliminated. Each remaining stream and lake was then buffered, using a GIS, to an arbitrary distance of 200–400m to produce a riparian zone map. Wetlands were either mapped for the first time from USGS 1:100,000 topographic maps or were extracted from USFWS wetland inventory maps. Wetlands and riparian zones were then combined with county and vegetation map layers within the GIS. Using this combined data layer and the species distribution databases, the ranges of species associated with wetland and riparian habitats was predicted within wetland and riparian zones. In this way distributions of animals, such as the common loon (*Gavia immer*) and river otter (*Lutra canadensis*), were predicted more realistically. Some species, such as the water shrew (*Sorex palustris*) and muskrat (*Ondatra zibethicus*), are associated with hydrographic features too small to be mapped at 1:100,000. Based on the assumption that adequate wetland and riparian microhabitats exist within each vegetation polygon, gap analysis continues to use simple county and vegetation data layers to predict the distribution of these microhabitat-dependent species.

Land ownership and management status Management practices and the range of management possibilities are related to land ownership. Thus gap analysis must assign attributes for both ownership and management. As ownership and management are so interlinked for these data to be useful, ownership and management information must be current.

Scott et al. divided management status into four categories:

1. an area, with an active management plan in operation, that is maintained in its natural state and within which natural disturbance events

are either allowed to proceed without interference or are mimicked through management. (This category includes most national parks, Nature Conservancy preserves, some wilderness areas, USFWS national wildlife refuges, and Audubon Society preserves.)

2. an area that is managed for its natural values but which may be subjected to uses that degrade the quality of natural communities that are present. (Most wilderness areas, USFWS refuges, and BLM areas of critical environmental concern are included in this class.)
3. an area for which legal mandates prevent permanent conversion to anthropogenic habitat types and that confer protection to federally listed endangered or threatened species. (Most designated public lands, including USFS, BLM, and state park lands are of this type.)
4. private or public land without an existing easement or irrevocable management agreement that maintains native species and natural communities, and which is managed primarily or exclusively for intensive human activity. (Urban, residential, and agricultural lands, public buildings and grounds, and transportation corridors are included in this category.)

Regionalization Scott et al. note that though gap analysis is being conducted by various parties primarily at the state level (the New England region is an exception), political boundaries rarely reflect biogeographic boundaries. To avoid biases or incomplete findings that are likely to result from confining gap analysis projects to the state level, the authors propose that GIS data layers be regionalized to generate species-richness and protection gap maps from the perspective of ecoregions rather than political units.

Limitations Scott et al. caution that gap analysis is an initial, coarse-scale approach to conservation evaluation and "not a panacea for conservation planners." They note that gap analysis as currently practiced has the following limitations:

- Vegetation maps derived from satellite imagery cannot show habitat smaller than the minimum mapping unit, thus critical microhabitats are missed or under-represented. Such microhabitats may be captured by subsequent higher resolution mapping efforts of key or priority areas identified by the initial coarse-scale analysis.

- Vegetation maps do not depict stand age, except for early successional stages of forests following clearcutting or fires. Gap analysis can identify relatively large areas of unfragmented forest, but at the present scale this method cannot indicate what proportion of the forest is old growth.
- Boundaries between vegetation types are in reality environmental gradients and not the sharp edges implied by the discrete polygons within a gap analysis GIS.
- Species distribution maps are only predictions based on known distributional limits and inferred habitat relationships. Nevertheless, gap analysis predictions have been shown to be reasonably accurate (70% or better) when compared to data from well-studied field sites.
- Maps of predicted habitat distribution do not reflect habitat quality or population density. Gap analysis predicts only the presence or absence of a species and not its rarity or abundance in a particular area (polygon).
- Gap analysis is not a substitute for endangered and threatened species listing and recovery efforts. It is, however, proactive—seeking to identify areas of critical conservation value that should be the foci of biodiversity management efforts.
- Gap analysis is not a substitute for a thorough national biological inventory. It does, however, provide a rapid assessment of the present distribution of vegetation and associated animal species, and can direct more intensive surveys to key or critical habitat. Stratified sampling based on gap analysis findings will make intensive field inventories considerably more cost effective in terms of time, labor, and money.
- Gap analysis is not the end point; it is merely the first step in what needs to be a comprehensive land conservation planning program. Gap analysis provides regional scale baseline data that can both guide more intensive surveys and allow changes in habitat availability and distribution to be monitored over time.
- Gap analysis relies on satellite remote sensing of vegetation and on knowledge of wildlife habitat relations to predict the distribution and current protection status of biodiversity. Scott et al. stress, however, that field investigations are essential for the development of specific management plans for any given area or region.

Benefits and Limitations of Remote Sensing Analyses

Interviews with the authors of the studies described in this chapter generated a wide range of comments as to the benefits and limitations of using remote sensing imagery.

Benefits

- ☺ Remote sensing imagery constitute an up-to-date and frequent source of information, as well as a twenty-year or longer time series of historical data that can be used to monitor spatial and temporal changes in land cover, land use, and green vegetation biomass.
- ☺ Relatively rapid regional-scale assessments of land cover, land use, and green vegetation biomass can be conducted semiautomatically; such would be impossible using field-based surveys.
- ☺ Powerful and relatively inexpensive computer hardware and software are now within reach of individuals.

Pitfalls to Avoid

- ☹ Results of remote sensing studies can be no more accurate than the user's knowledge of the land use history and spectral characteristics of the area.
- ☹ "Push-button" image processing software often gives the first-time user the erroneous impression that generating a classification of a given area is an automated, almost instantaneous process. In reality, to transform a spectral classification into an accurate land use or land-cover classification, the user must have considerable knowledge of the area, the characteristic reflectances of soil, vegetation, and water, and the distorting effects of the atmosphere, sun angle, and landscape relief and aspect. Creating a spectral classification is indeed easy; but relabeling the spectral classes to reflect land-cover classes of interest requires time and knowledge.
- ☹ Cloud and haze cover can severely limit access to high-quality imagery during certain seasons in temperate areas, and can preclude image acquisition in the tropics.
- ☹ The interaction of topographic relief with low sun angle (associated in high latitudes with winter, and everywhere with early morning over-

pass times of the remote sensing satellite) can result in very heavy shadowing and sensor inability to distinguish important land-cover types.

- ☹ Accurate geometric correction of remote sensing data can be very difficult in wilderness areas with little human activity (no roads, railways, etc.) because of few unambiguously identifiable ground control points, and sometimes because of a lack of accurate large-scale topographic maps. Access to inexpensive, hand-held Global Positioning System receivers has helped to improve this situation.
- ☹ The timing of image acquisition relative to field surveys is very important, as landscapes are not static. Landscapes change in response to weather patterns, the seasonal or super-annual growth and senescence of vegetation, and unpredictable human and natural disturbances.
- ☹ The trade-off that all remote sensing systems make between area coverage and spatial resolution often means that available imagery have neither the most appropriate spatial resolution nor the best field of view for addressing the user's research question. Because coarse-scale imagery tends to be less expensive than high spatial resolution imagery (excluding aerial videography), cost considerations may also force the user to rely on imagery that is not ideal.
- ☹ Too often, organizations and agencies purchase high-end hardware and software for image processing and for geographic information display and management without first assessing whether it offers researchers or managers a tool that they are interested in using. Learning to use this tool is time consuming, and it is most easily done when a particular problem can be solved, realistically, only by remote sensing analysis. First time users unsure of the long-term utility of remote sensing image analysis should opt for low-cost, but expandable, hardware and a shallow learning curve software. If the tool proves useful, then both hardware and software can be upgraded when necessary.

Appendix: Online Searches for and Viewing of Remote Sensing Imagery

NOAA AVHRR Image Archive

OSCAR is an on-line catalog of archived imagery accessible

- directly by modem (1-800-528-2415; set modem to full-duplex, VT100, no parity 8 data bits, 1 stop bit, baud 1200 to 9600), or
- via the Internet, using Telnet (telnet 140.90.110.5).

Global Land Information System (GLIS)

GLIS is an on-line land data directory, guide, and inventory system developed by the U.S. Geological Survey at EROS data center in Sioux Falls, South Dakota. GLIS was designed to provide a single interactive source of information about, and access to, data pertaining to the earth's land surface. GLIS is accessible in several ways:

- via an ASCII text menu system via modem (tel: 605-594-6888; 8 bits, no parity, 1 stop bit) or Internet (telnet glis.cr.usgs.gov).
- via a PC graphical interface that allows: interactive selection of geographic areas of interest, the graphical presentation of each inventory item over a base map, and viewing of a raster browse image that has been downloaded to the user's PC. PC-GLIS requires an 80286 or greater, DOS 3.0+, a Hayes compatible modem, VGA, and 1.5 Mb of disk space for software.

PC-GLIS can be obtained by

- calling 1-800-252-4547,
- e-mail request to glis@glis.cr.usgs.gov,
- mail request to GLIS User Assistance, USGS, Eros Data Center, Sioux Falls, SD 57198, USA, or
- via the Internet: FTP edcftp.cr.usgs.gov; log in as "anonymous" with your e-mail address as the password; cd /pub/software/pcglis . . . change to pcglis directory; binary (set system to binary mode); mget * (downloads two files README [install instructions] and PCGLIS.EXE),
- via the World Wide Web: point your web browser (Lynx, Mosaic, Netscape, etc.) to http://edcwww.cr.usgs.gov/glis/glis.html.

GLIS is also accessible via an X terminal interface that features: graphical area coverage, graphic coverage display, inventory browse, and guide figures. The X terminal interface requires: an X terminal emulator running X11 release 4 or later, and a direct connection to the Internet (modems, even 28.8 k baud, are too slow). Connections to XGLIS are by starting the X-terminal emulator, beginning a telnet session (telnet xglis.cr.usgs.-gov), and setting the display to the name of your X-terminal.

Examples of GLIS metadatasets are AVHRR imagery, USGS land use and land-cover data, North American Landscape Characterization (NALC) multitemporal MSS datasets, National Aerial Photography Program (NAPP) air photos, Landsat MSS imagery, and sources for DEM, and vector maps such as the Digital Chart of the World.

NOAA Satellite Active Archive

NOAA's satellite active archive (SAA) enables users on the Internet to quickly search, browse, order, and receive NOAA AVHRR satellite data by mail or over the network. SAA uses a text or graphical (X-Windows) interface that mimics GLIS. SAA can be accessed via Telnet (telnet saa. noaa.gov, or telnet 140-90.232.101). SAA is also a node on the World Wide Web (http://www.saa.noaa.gov/) and can be accessed using Lynx or Mosaic (two hypermedia software packages for navigating the Internet).

Imagery Available Through GOPHER, FTP and the World Wide Web on the Internet

Imagery in GIF and JPEG formats are available on various GOPHER, FTP and World Wide Web servers on the Internet. For example, the Australian Environmental Resources Information Network (ERIN) provides access to NDVI/AVHRR pictures of Australia (point web browser to http://kaos.erin.gov.au/). NASA Gophers and FTP servers provide access to pictures from a variety of photographic and satellite image sources (FTP images.jsc. nasa.gov). A list of "Gopher Jewels" is available through gopher or directly from the compiler, David Riggins, at RIGGINS_DW@dir.texas.gov.

NASA has committed itself to providing its satellite imagery over the Internet. The Remote Sensing Public Access Center (RSPAC) promotes improved public access to NASA's huge repository of earth (and space) imagery. RSPAC can be reached at http://www.rspac.ivv.nasa.gov/ and NASA's home page is at http://www.gsfc.nasa.gov/NASA.homepage .html.

In 1994, NSF, ARPA and NASA jointly sponsored the Alexandria Digital Library at the University of California, Santa Barbara. The purpose of the Alexandria Digital Library is to design and implement a digital library for spatially indexed information. The Alexandria Digital Library has a home page on the Web at http://alexandria.sdc.ucsb.edu/.

World Wide Web addresses change from time to time, and traffic on the Web has caused many popular sites to become saturated. If you can't get through to a site, try again later. If you think a site has changed its address, use one of the search engines on the Web (e.g., WebCrawler) to find the new address. An updated set of pointers to all of the sites mentioned in this book will be available at the home page for this book at http:/bandersnatch.fnr.umass.edu/pub/rs.html.

Glossary

active remote sensing A system based on the illumination of a scene by artificial radiation and the collection of the reflected energy returned to the system. Examples are radar and systems using lasers. *Compare with passive remote sensing.*

acuity A measure of human ability to perceive spatial variations in a scene. It varies with the spatial frequency, shape, and contrast of the variations, and it depends on whether the scene is color or monochrome.

additive primary colors The spectral colors red, green, and blue can be used to create all other colors when mixed by projection through filters. This discovery was made by Thomas Young (1773–1829). *See RGB; compare with subtractive primary colors.*

aerial videography *See videography.*

analog data A form of data in which values are continuous (not discrete). Analog data are displayed in graphic form, such as curves.

analog image An image in which the continuous variation in the property being sensed is represented by a continuous variation in image tone. In a photograph this is achieved directly by the grains of photosensitive chemicals in the film; for an electronic detector the response (in, say, millivolts) is transformed to a video display where it may be photographed. *Compare with digital image.*

ancillary data Secondary data pertaining to the area or classes of interest, such as topographical, demographic, or climatological data. Ancillary data may be digitized and used in the analysis in conjunction with the primary remote sensing data.

angle of depression In SLAR usage, the angle between the horizontal plane passing through the antenna and the line connecting the antenna and the target.

angle of incidence 1. The angle between the direction of incoming EMR and the normal to the intercepting surface. 2. In SLAR systems this is the angle between the vertical and a line connecting antenna and target.

angle of reflection The angle that EMR reflected from a surface makes with the perpendicular (normal) to the surface.

atmospheric windows A range of wavelength in which radiation can pass through the atmosphere with relatively little attenuation.

AVHRR Advanced Very High Resolution Radiometer. Multispectral imaging system carried by the TIROS-NOAA series of meteorological satellites.

AVIRIS Advanced Visible Infrared Imaging Spectrometer. A 224 channel airborne multispectral scanner with contiguous bands between 0.4 and 2.45 μm.

azimuth The geographical orientation of a line given in degrees clockwise from north. In radar terminology it refers to the direction at right angles to the radar pulse direction, which is parallel to the ground track in a sideways-looking system.

backscatter The scattering of radiant energy from a feature back toward the detector.

band The range of wavelengths from which data are gathered by a recording device.

bilinear interpolation Mathematically determining the value for a pixel based on the value of its four nearest neighbors.

bits An abbreviation of binary digits; bits are numbered as the exponent of 2 (8 bits = 2^8, or 256 binary choices, 0–255).

blackbody A perfect radiator and absorber of electromagnetic energy, meaning that all incident energy is absorbed. No natural material behaves as a true blackbody, although platinum black and other soots closely approximate this ideal.

brightness The visual perception that an area appears to emit more or less light. It is a measure of the EMR reflected, transmitted, or emitted from an area.

byte Eight bits of digital data, representing numbers between 0 and 255.

CAMS Calibrated Airborne Multispectral Scanner. An experimental airborne multispectral scanner developed by NASA.

CCD charge-coupled device. A light-sensitive solid-state device that generates a voltage proportional to the intensity of illumination. This device can be charged and discharged very quickly, and it is used in pushbroom devices, spectroradiometers, and modern video cameras.

change-detection images Images generated by comparing two images acquired at different times.

class A terrain feature that is of interest to the researcher, such as forest type or water condition.

classification The process of assigning individual pixels of a multispectral image to categories based on spectral reflectance characteristics.

cluster analysis The statistical analysis of pixels in a multispectral image to detect their inherent tendency to form clusters in multidimensional brightness space.

CMY (CMYK) cyan-magenta-yellow-black. The primary subtractive colors used in printing to generate the full range of visible colors. *Compare with RGB.*

color The characteristic of an object that is dependent on the wavelength of the light it reflects. If the light is of a single wavelength, the color seen will be a pure spectral color. If light is composed of two or more wavelengths, the color will be mixed.

color composite A color image produced by assigning a color to a particular spectral band. In Landsat MSS blue is often assigned to band 4, green to band 5, and red to band 7 to produce an image that approximates the looks of a color infrared photograph.

color infrared film Photographic film sensitive to energy in the visible and near-infrared wavelengths, generally from 0.4–0.9μm; usually used with a minus-blue (yellow) filter, which results in an effective film sensitivity of 0.5–0.9μm. Color infrared film is especially useful for detecting changes in the condition of vegetation. Color infrared film is not sensitive to the thermal infrared region and therefore cannot be used as a heat-sensitive detector.

complex dielectric constant An electrical property of matter that influences radar returns.

contrast Ratio of the energy emitted or reflected by an object and the surrounding background.

contrast stretching Expanding a range of brightness values in an image to utilize the full contrast range of the recording film or display device. Increasing the contrast within an image helps to highlight differences between features.

convolution Mathematically determining the value for a pixel based on an arithmetic combination of its neighbors. *See interpolation, nearest neighbor, bilinear interpolation, and cubic convolution.*

coordinates, geographical Pairs of numbers (e.g., latitude-longitude; northing-easting) for describing the positions of points on the earth.

coregistration Overlaying two images such that each pixel in image 1 matches exactly the position of the corresponding pixel in image 2.

corner reflector Terrain features intersecting at right angles; radar is thus reflected directly back to its source.

CRT cathode ray tube. The monochrome or color display device used in televisions and computer screens.

CSIRO Commonwealth Scientific and Industrial Research Organization, Australia.

cubic convolution Mathematically determining the value for a pixel based on the value of its sixteen nearest neighbors.

cursor An aiming device, such as a lens with crosshairs on a digitizer or a pointer on computer display.

CZCS Coastal Zone Color Scanner. A multispectral imaging system carried by the Nimbus series of meteorological satellites.

decision rule (or classification rule) The criterion used for classification (e.g., parallelepiped, minimum distance to means rule, maximum likelihood rule).

decorrelation stretch Use of principle components analysis to enhance the contrast of digital imagery by removing the correlation between spectral bands. *See principal components analysis.*

density slicing The process of converting the full gray tone range of data into a series of intervals, or slices—each of which expresses a range in the data.

depression angle The angle between the horizontal plane passing through the radar antenna and the line connecting the antenna to the target. It is easily confused with the *look angle.*

detection A feature is detected if the sensor is able to assign to it a brightness value that contrasts with the surrounding terrain. Detection of a feature does not imply that the researcher is able to identify the feature as belonging to a particular land-cover class.

detector (of radiation) A device providing electrical output that is a useful measure of incident radiation.

diffuse reflector Any surface that reflects incident radiation in many directions; it is thus the opposite of a *specular reflector.* Cloth is a good example of a diffuse reflector, whereas mirrors are examples of specular reflectors in the visible portion of the electromagnetic spectrum. Almost all terrain features (except calm water) act as diffuse reflectors of incident solar radiation. The smoothness or roughness of a surface depends on the wavelength of the incident EMR.

digital image An image for which the incident radiation recorded as a voltage by a detector has been converted from a continuous range of analog values to a range expressed by a finite number of integers—for example, recorded as binary codes from 0 to 255 (8 bits). *Compare with analog image.*

digitization The process of converting an image originally recorded on photographic material into numerical raster or vector format.

Doppler shift A change in the observed frequency of electromagnetic or other waves caused by the relative motion between source and detector. It is used principally in the generation of *synthetic-aperture radar* images.

edge The boundary of an object in a photograph or image. An edge is characterized by an easily detectable change in the gray tone value at the border of two terrain features.

edge enhancement The mathematical process of increasing the contrast between adjacent areas with different tones (brightnesses) within an image. *Compare with smoothing.*

electromagnetic spectrum The continuum of electromagnetic radiation (EMR) extending from the shortest cosmic rays, through gamma rays, X rays, ultraviolet radiation, visible radiation, infrared radiation, and including microwave and all other wavelengths of radio energy.

emission The process by which a body emits EMR, usually as a consequence of its temperature alone.

emissivity The ratio of the energy radiated by a material to that which would be radiated by a blackbody at the same temperature. A blackbody has an emissivity of 1 and natural materials range from 0 to less than 1.

EMR electromagnetic radiation. Energy propagated through space or through material. The term *radiation,* alone, is commonly used for this type of energy.

EOSAT Earth Observation Satellite Company. A private company based in Lanham, Maryland, which since September 1985 has been contracted by the U.S.

government to market Landsat data and to develop replacements for the Landsat system.

ERDAS Earth Resources Data Analysis System. A map preparation and image processing software package, now called ERDAS Imagine, produced by ERDAS, Inc.

EROS Earth Resources Observation System. Administered by the U.S. Geological Survey at the EROS Data Center, Sioux Falls, South Dakota. It forms an important source of image data from Landsat 1, 2, and 3 and for Landsat 4 and 5 data up to September 1985, as well as for airborne data covering the United States.

ESA European Space Agency. Owned by a consortium of several European states, this agency is dedicated to the development of space science, including the launch of remote sensing satellites.

false color The use of one color to represent another; for example, red may be chosen to represent nonvisible infrared radiation.

false color image A color image in which parts of the nonvisible electromagnetic spectrum are expressed as one or more of the red, green, and blue components, so that the colors produced by the terrain do not correspond to human visual experience. In multispectral imagery it is also called a false color composite (FCC). The most commonly seen false color images display the very near infrared as red, red as green, and green as blue.

feature A terrain object that exhibits distinctive radiation brightness, texture, shape, or context.

feature extraction Characterization of the multispectral signature of terrain features of interest to the researcher.

field of view The swath of terrain visible to a sensor and which is determined by the fixed angle through which a sensor is exposed to radiation and the altitude of the sensor platform. *See* IFOV.

filtering Removal of certain spectral or spatial information, using physical (glass) or software filters, to highlight or diminish features in the resulting image.

flight line A line drawn on a map or image that represents the track over which an aircraft has been flown or is to fly.

GAC global area coverage. The 4km spatial resolution imagery generated by the NOAA AVHRR sensor.

gain An increase in signal power.

GCP ground control point. A point in two dimensions that is recognizable on both an image and a topographic map, and can be represented by (x, y) coordinates based on the map's cartographic projection and grid system. It is used to geometrically correct an image to a standard map projection.

geographical (geometrical) correction Adjustments made in an image to change its geometrical character, so that it conforms to a standard map projection such as Universal Transverse Mercator.

geostationary (geosynchronous) orbit A west-to-east equatorial orbit at an altitude of 36,000km that matches the speed and direction of the earth's rotation so that a satellite remains over a fixed point on the earth's surface.

GEOS Geostationary Operational Environmental Satellite. A series of geostationary satellites that comprise part of a global weather monitoring system. The GEOS satellites are the platforms for the Visible Infrared Spin-Scan Radiometer sensor.

GIS geographic information system. A data management, manipulation, analysis, and display system based on sets of data that combine the characteristics of an object or feature and its geographical location. The datasets may be map oriented, where features are represented in vector format by lines, points, and polygons, or image oriented, where features are represented in raster format as integer value cells in a rectangular grid.

GMS Geostationary Meteorological Satellite. A Japanese weather satellite that is one of the five satellites in the global geostationary weather satellite network.

GOES Geostationary Operational Environmental Satellite. A U.S. weather satellite that is one of the five satellites in the global geostationary weather satellite network.

GPS Global Positioning System. A global network of 24 radio transmitting **NAVSTAR** (Navigation System with Time And Ranging) satellites developed by the U.S. Department of Defense to provide accurate navigation and geographic location 24 hours a day anywhere on the globe. The NAVSTAR satellites circle the earth in 20,200km, circular orbits with a 12-hour period. The orbital geometry of six 55° inclined planes with three satellites in each plane (six satellites are used as "spares") will enable reception of direct line-of-sight navigation signals from at least four satellites at any point at or near the earth's surface at all times. Each satellite transmits a coarse/acquisition navigation signal that provides users with location accuracy ranging from 15m using one GPS receiver to a centimeter or less when two GPS receivers are "connected" in differential mode. Simultaneous monitoring of three satellites gives two-dimensional (latitude and longitude) position when altitude is known. Four satellites provide complete three-dimensional positioning.

gray scale A monochrome continuum of shades ranging from white to black with intermediate shades of gray.

ground resolution cell The area on the terrain that is covered by the instantaneous field of view of a detector. The size of the ground resolution cell is determined by the altitude of the remote sensing system and the instantaneous field of view of the detector.

ground truth, ground data Slang used for data and information obtained about terrain features to aid in interpretation of remotely sensed data. These data should be gathered, if possible, concurrently with the acquisition of remote sensing imagery. Data on weather, soils, and vegetation types and conditions are typically gathered.

hardcopy device A printer, plotter, laser writer, or CRT imager that creates a paper, film, or slide facsimile of an image.

HDTV High Definition Television. A new technology that will substantially increase the resolution (number of lines) that can be displayed on television screens.

HSI The hue, saturation, and intensity color model used by color televisions and videorecorders to display all colors in the visible spectrum. A particular HSI color is defined by its location on a double cone with greatest circumference at the midpoint of the central (z) axis. Moving round the circumference changes the hue, saturation decreases inward toward the z axis of the cone, and intensity changes by moving up and down the z axis. Gray tones are generated by moving along the z axis.

histogram A graphical means of expressing the frequency of occurrence of values as bars of equal range aligned along the horizontal axis. The height of each bar (along the vertical axis) represents the frequency at which values in the dataset fall within the chosen range.

HRV high resolution visible sensor. A three-waveband (green, red, and near-IR) multispectral sensor with 20m spatial resolution deployed on the SPOT satellite system.

hue The color of a feature that distinguishes it from gray of the same brightness such that it can be characterized as red, yellow, green, blue, or intermediate shades of these colors.

IFOV instantaneous field of view. The patch of terrain visible to a detector at any one time, determined by the fixed angle through which a sensor is exposed to radiation and the altitude of the sensor platform. The minimum IFOV is constrained by the intensity and duration of the radiation falling on a detector; it constitutes one limit to the resolution of a remote sensing system.

image A recorded representation of an object produced by optical or electronic means. It most often refers to when the EMR emitted or reflected from a scene is not directly recorded on film.

image enhancement Any operation that improves the detectability of features of interest. These operations include contrast improvement, edge enhancement, spatial filtering, noise suppression, image smoothing, and image sharpening.

image processing All the various operations that can be applied to photographic or image data.

image striping A defect in across-track, spin, and pushbroom scanners produced by the nonuniform response of a single detector or array of detectors. The stripes are perpendicular to flight direction in an across-track scanned image, but parallel to it in a pushbroom image.

incidence angle The angle between the surface and an incident radar signal.

intensity A measure of the energy reflected or emitted by a surface.

interpolation Mathematically determining the value for a pixel based on an arithmetic combination of its neighbors. Filtering, resampling, and geometric correction all use interpolation. Interpolation based on the mean of neighborhood pixels should not be undertaken on images where the mean is undefined (i.e., categorical data). The most common interpolation techniques are nearest neighbor, bilinear interpolation, and cubic convolution. *See convolution.*

IR (infrared) Energy in the 0.7–300μm wavelength region of the electromagnetic spectrum. For remote sensing the infrared wavelengths are subdivided into near infrared (0.7–1.3μm), middle infrared (1.3–3.0μm), and far infrared (>

3.0μm). The far infrared range of 3.0–15.0μm is referred to as *thermal* or *emissive infrared.*

Kauth-Thomas Tasseled Cap A set of linear transformations that converts Landsat MSS and TM multispectral imagery to represent the greenness, yellowness, and wetness within the terrain.

Kelvin A thermometer scale starting at absolute zero (-273°C) and having degrees of the same magnitude as those of the Celsius thermometer. Thus 0°C = −273°K; 100°C = 373°K.

LAC local area coverage. The 1km spatial resolution imagery generated by the NOAA AVHRR sensor.

LAI leaf area index. A unit-less measure of the ratio of the surface area of all leaves to the ground area.

Landsat A series of remote sensing satellites in sun-synchronous, polar orbit, with first launch in 1972. The Landsat satellites are the platforms for the MSS, RBV, and TM sensors. The Landsat program was initially administered by NASA, then NOAA, and since 1985 by EOSAT. The Land Remote Sensing Policy Act of 1992 (Public Law 10255) will return administration of Landsat to NASA, in conjunction with the Department of Defense, by 1997.

LFC large format camera. A large format analog camera used experimentally on board the space shuttle to obtain high resolution (5–10m) remote sensing images.

light Visible radiation, about 0.4–0.7μm in wavelength.

line scanner An imaging device that uses a mirror to sweep the ground surface below the flight path of the platform. An image is built up as strips of pixels.

look angle The angle between the vertical plane containing a radar antenna and the direction of radar signal. It is complementary to the *depression angle.*

look direction The direction in which pulses of radar are transmitted.

map A representation of the physical features (natural, artificial, or both) of a part of the earth's surface, in two dimensions, at an established scale, with the means of orientation indicated.

map projections Cartographers must decide which map characteristic (the area, shape, scale, or direction) is to be depicted accurately at the expense of others, as a curved surface cannot be projected onto a flat surface without distortion. Cartographers use a variety of map projections (Albers equal area, Universal Transverse Mercator, Peter's) to depict the curved surfaces of the globe as flat, two-dimensional maps.

maximum likelihood A statistical decision rule used in supervised classification to assign pixels to the class of highest probability according to their spectral characteristics.

median filter A spatial filter that substitutes a pixel's brightness value with the median brightness value from surrounding pixels. It is useful for removing random noise.

METEOSAT A European weather satellite that is one of the five satellites in the global geostationary weather satellite network.

micrometer (μm) A unit of length equal to one-millionth of a meter or one thousandth of a millimeter.

microwave Electromagnetic radiation having wavelengths between 1m and 1mm, or 300–0.3 GHz in frequency. It is bounded on the short wavelength side (1mm) by the far infrared and on the long wavelength side by very high frequency radio waves. Passive systems operating at these wavelengths are sometimes called microwave systems. Active systems are called radar, although the literal definition of radar requires a distance-measuring capability not always included in active systems.

Mie scattering The scattering of electromagnetic energy by particles in the atmosphere with comparable dimensions to the wavelength involved.

minimum distance to centroid classifier A method of assigning pixels within an image to thematic classes during a supervised classification. Pixels within training sites are used to calculate mean vectors that constitute the center location (centroid) of each class in *n*-dimensional brightness space. For each pixel the classifier calculates the Euclidean distance to each class centroid and assigns the pixel to the class with the shortest pixel-to-centroid distance. The minimum distance to centroid (means) logic of this classifier allows all pixels to be assigned unambiguously to one class. Such complete classification is possible because, unlike regular-sided parallelepipeds, the *n*-dimensional polygons that surround each centroid are irregularly shaped and their boundaries are contiguous—leaving no empty, unclassified space.

MIPS, Map and Image Processing System A map preparation and image processing software package, now called TNT-Mips, produced by MicroImages, Inc.

mosaic A set of overlapping aerial photographs or images whose edges have been matched to form a continuous pictorial representation of a portion of the earth's surface.

mosaicking Assembling photographs or other images to form a continuous representation of a portion of the earth's surface.

MSS multispectral scanner. An across-track imaging device carried by Landsats 1 through 5, which records scenes in four wavebands (three in the visible and one near-IR) with a spatial resolution of 79m.

multispectral sensor A system for recording the brightness of the terrain within several spectral bands simultaneously.

nadir The point on the ground vertically beneath the sensor.

NASA National Aeronautics and Space Administration (U.S.).

NDVI normalized difference vegetation index. An index of vegetation biomass computed by dividing the difference of the near-ir and visible red bands (Landsat MSS bands 7 and 5, Landsat TM bands 4 and 3, AVHRR bands 2 and 1) by their sum.

near-polar orbit An orbit that passes close to the poles, thereby enabling a satellite to pass over most of earth's surface, except the immediate vicinity of the poles themselves.

nearest neighbor Mathematically determining the value for a pixel based on the value of its nearest neighbor.

NOAA National Oceanic and Atmospheric Administration (U.S.).

noise Random or regular interfering effects that degrade the information quality within an image. Noise is due to defects in the recording device.

nonselective scattering The scattering of electromagnetic energy by particles in the atmosphere which are much larger than the wavelengths of the energy, thus causing all wavelengths to be scattered equally.

nonspectral hue A hue that is not present in the spectrum of colors produced by the splitting of white light by a prism or diffraction grating. Examples are brown, magenta, and pastel shades.

NTSC National Television Standards Committee video. The video standard format for the United States.

orbit The path of a satellite around the earth under the influence of gravity.

orthophotograph A vertical aerial photograph from which the distortions due to varying elevation, tilts, and surface topography have been removed, so that it represents every object as if viewed directly from above, as in a map.

PAL phase alternating line. The video standard used almost everywhere in the world outside the United States.

panchromatic Films or digital images that record terrain brightness across a broad range of EMR wavelengths (e.g., entire visible part of spectrum) and which are displayed as a grayscale image.

parallelepiped classifier A simple and fast method of assigning pixels within an image to thematic classes during a supervised classification. Maximum and minimum brightness values of pixels within training sets are used to define class boundaries which can be envisioned as n-dimensional prisms. As class boundaries often overlap one another, and when combined do not fill all the n-dimensional space, pixels can ambiguously be members of more than one class, or can remain unclassified.

passive microwaves Radiation in the 1mm to 1m range emitted naturally by all materials.

passive remote sensing Recording images that represent the reflection or emission of electromagnetic radiation that has a natural source. *Compare with active remote sensing.*

PCI EASI/PACE A map preparation and image processing software package, produced by PCI Remote Sensing Corp.

photogrammetry Photographic methods for obtaining reliable measurements of features of interest.

photograph A picture formed by the action of light on a base material coated with light-sensitive chemicals. *Compare with image.*

photographic (photo) interpretation Examining photographic images to identify and measure features of interest.

pixel "picture element." A datum in a digital image, having both a spatial aspect (its position in the image and its rectangular dimension on the ground) and a spectral aspect (the brightness in a particular waveband of the terrain within the pixel's rectangular dimension on the ground).

polarization The direction of vibration of the electrical field of electromagnetic radiation. In SLAR systems polarization is either horizontal or vertical.

principal components analysis, PCA A statistical technique for reducing the number of bands in an image by generating a series of components, each of which is an uncorrelated combination of the raw bands. Each successive component in a PCA contains less variance, and as such the first few components will contain most of the information found within all the raw bands.

pulse A short burst of EMR transmitted by the radar.

pushbroom system An imaging device consisting of a fixed linear array of many detectors (*see* CCD) that records terrain brightness of a line of pixels at one time. Each line of data is aligned perpendicular to the motion of the platform.

radar The acronym for "radio detection and ranging." Remote sensing by radar uses pulses of artificial electromagnetic radiation in the 1mm to 1m range to map objects that reflect the radiation. The position of the object in the resulting image is a function of the time that a pulse takes to reach it and return to the antenna.

radiation The emission and propagation of energy through space or through a material medium in the form of waves, e.g., the emission and propagation of electromagnetic waves or of sound waves. The process of emitting radiant energy.

radiometer An instrument for quantitatively measuring the intensity of EMR in some band of wavelengths in any part of the electromagnetic spectrum.

radiometric correction Correcting gain and offset variations in the radiation data generated by sensor detectors.

range The distance in the direction of a radar signal, usually to the side of the platform in an imaging radar system. The slant range is the direct distance from the antenna to the object, whereas the ground range is the distance from the ground track of the platform to the object.

raster format Representation of a surface as a grid of pixels. *Compare with vector format.*

Rayleigh scattering Selective scattering of light in the atmosphere by particles that are small compared with the wavelength of light.

RBV Return-Beam Vidicon. A multispectral framing sensor system aboard Landsats 1, 2, and 3.

real-aperture radar An imaging radar system used in SLAR systems, where the azimuth resolution is determined by the physical length of the antenna, the wavelength, and the range of the feature. Return signal timing and strength are used directly to generate the final image. *Compare with synthetic-aperture radar.*

rectification *See geometrical correction.*

reflectance The ratio of the electromagnetic energy reflected by a surface to that which falls upon it; it is often associated with the prefix "spectral."

reflection The EMR that is neither absorbed nor transmitted. Reflection may be diffuse, when the incident radiation is scattered upon being reflected from the surface; or it may be specular, when all or most angles of reflection equal the angle of incidence.

registration The process of geometrically aligning two or more images such that pixels representing a single ground area can be digitally or visually superimposed. Data being registered may be of the same type, from very different kinds of sensors, or collected at different times.

remote sensing The acquisition of information on an object or phenomenon by a recording device that is not in physical or intimate contact with the object or phenomenon under study. The technique employs such devices as cameras, radar or sonar systems, radiometers, seismographs, gravimeters, magnetometers, and scintillation counters.

resampling The calculation of new brightness values for pixels created during geometric correction of a digital scene based on the values in the local area around the uncorrected pixels.

resolution The spatial, spectral, radiometric, and temporal detail possible for a given sensor system.

RGB The red, green, blue color model used by most computer displays to generate all colors of the visible spectrum. In an 8 bit computer display the range of each color is from 0 to 255. The RGB combination of 0,0,0 = black; 255,255,255 = white; 50,50,50 = dark gray; 200,200,200 = light gray; 255,0,0 = bright red; 50,0,0 = dark red; 200,0,200 = purple. *Compare with* CMY.

sampling rate The temporal, spatial, or spectral rate at which measurements of physical quantities are taken. Temporally, sampling variables may describe how often data are collected or the rate at which an analog signal is sampled for conversion to digital format; the spatial sampling rate describes the number, ground size, and position of areas where spectral measurements are made; the spectral sampling rate refers to the location and width of the sensor's spectral channels with respect to the electromagnetic spectrum.

saturation The maximum value that can be assigned to a pixel in a digital image, regardless of its actual brightness. In color theory it means the degree of mixture of a pure hue and neutral gray.

scale The ratio of the size of a feature on a map, photograph, or digital image to the size of the same feature on the ground. Scale may be expressed as a ratio (1:24,000), a representative fraction (1/24,000), or an equivalence (1 in = 2,000 ft). The larger the representative fraction, the larger is the scale of the map or image (i.e., 1/100 is large-scale, whereas 1/1,000,000 is small-scale).

scan line The narrow strip on the ground that is swept by the instantaneous field of view of a detector in a scanner system.

scanner A sensor that uses an oscillating or spinning mirror to focus terrain brightness onto a single detector or array of detectors.

scattering An atmospheric effect in which electromagnetic radiation, usually of short visible wavelength, is propagated in all directions by encounters with gas molecules and aerosols. *See Rayleigh, Mie, and nonselective scattering.*

scene The area on the ground recorded by a photograph or other image, including the atmospheric effects on the radiation as it passes from its source to the ground and back to the sensor.

SECAM sequential coleur à memoire. The video standard used in France and Russia.

sensitivity The degree to which a detector responds to electromagnetic energy incident upon it.

sensor Any device that gathers energy (EMR or other), converts it into a signal, and presents it in a form suitable for obtaining information about the environment.

signal-to-noise ratio (S/N) The ratio of the level of the signal carrying real information to that carrying spurious information as a result of defects in the system.

signature The spectral properties of a feature, expressed as the range of brightness values in a number of spectral bands, that distinguish it from other features in the image.

SLAR side-looking airborne radar. Airborne systems that use *real-aperture radar* to generate images of the terrain.

smoothing The averaging of brightness values in adjacent pixels to produce more gradual transitions across the image. *Compare with edge enhancement.*

SMS-USA Synchronous Meteorological Satellite. A U.S. weather satellite that is one of the five satellites in the global geostationary weather satellite network.

spatial filter An image transformation, usually a one-to-one operator used to lessen noise or enhance certain characteristics of the image. Spatial filters use moving odd-sided square templates (grids) to reassign the brightness values of image pixels. The brightness transformation is usually a mathematical function of the brightness value of surrounding pixels combined with the value of the overlying cells within the template.

spectral band A range in the electromagnetic spectrum between two wavelengths or frequencies.

spectral reflectance Reflectance of electromagnetic energy by a feature at specific wavelengths or wavebands.

spectral regions Ranges of wavelengths subdividing the electromagnetic spectrum (e.g., the visible region, x-ray region, infrared region, microwave region).

spectral response The measurable energy reflected from and emitted by a feature.

spectral signature *See signature.*

spectrometer, or spectroradiometer A device to measure the spectral distribution of EMR reflected or emitted by a feature.

specular reflector Any object that reflects electromagnetic energy without scattering or diffusion. The surface of a specular reflector is smooth in relation to the wavelengths of incident energy. *Compare with diffuse reflector.*

SPOT Satellite Pour l'Observation de la Terre. A French satellite carrying two pushbroom imaging systems: (1) the three-waveband (green, red, and near-IR) multispectral high resolution visible (HRV) sensor, with 20m spatial resolution, and (2) the panchromatic sensor (visible spectrum only), with 10m resolution. The sensors are pointable so that off-nadir images are possible, allowing stereoimages to be generated.

Stefan-Boltzmann Law A radiation law stating that the energy radiated by a blackbody is proportional to the fourth power of its absolute temperature.

subtractive primary colors Hues created by using the light-absorbing (subtracting) characteristics of the three primary subtractive colors: cyan, magenta, and yellow. When all three colors are combined in equal quantities, all light is subtracted; the image thus appears to our eyes as black. Selective absorption (subtraction) of colors according to the different proportions of the three primary colors allows all colors to be generated. *Compare with additive primary colors.*

sun-synchronous orbit A near-polar orbit in which the satellite always crosses the equator at the same local solar time. Landsat and SPOT are examples of sun-synchronous orbiting satellites.

supervised classification A method of generating a thematic map from digital remote sensing imagery. Thematic classes are defined by the spectral characteristics of pixels within an image that correspond to training sets chosen to represent known features on the ground. Each pixel within the image is then assigned to be a member of a thematic class using one of several decision rules (parallelepiped, minimum distance, maximum likelihood). Supervised classification generates a thematic map directly from digital imagery. *Compare with unsupervised classification.*

swath width The linear ground distance in the across-track direction within which terrain brightness is recorded by the sensor system.

synoptic view The ability to image widely dispersed areas at the same time and under the same conditions.

synthetic-aperture radar (SAR) A radar imaging system in which high resolution in the azimuth direction is achieved by using the Doppler shift of backscattered waves to identify waves from ahead of and behind the platform, thereby simulating a very long antenna. *Compare with real-aperture radar.*

thematic map A map designed to highlight particular features or concepts (e.g., soils, vegetation types, land use). Topographical maps are not usually considered thematic maps.

thermal infrared The preferred term for the wavelength range of the IR region that extends roughly from 3μm at the end of the mid infrared to about 15 or 20μm, where the far infrared begins.

threshold The boundary in spectral space beyond which a pixel has such a low probability of inclusion in a given class that the pixel is excluded from that class.

TIMS Thermal Infrared Multispectral Scanner. An aircraft platform–based device developed and used by NASA to measure and create images in six wavebands of thermal IR emitted by the surface.

TM Thematic Mapper. An imaging device carried by Landsats 4 and 5, which records scenes in seven wavebands: six in the visible and near-IR (with a resolution of 30m) and one in the thermal-IR (with a resolution of 120m).

tone Each distinguishable shade of gray, from black to white, on a monochrome image.

training set, training site A sample of the earth's surface that represents a known feature of interest. The spectral characteristics (described statistically) of each training set within an image are used to determine class boundaries and pixel assignments in supervised classification.

unsupervised classification Generating a thematic map from digital remote sensing imagery by first statistically clustering pixels into classes according to their spectral similarity, then using the researcher's knowledge of the area to label spectral classes as features of interest. *Compare with supervised classification.*

USFWS United States Fish and Wildlife Service.

VCR video cassette recorder. A device that records/reads NTSC, PAL, or SECAM video signals on/from magnetic tape.

vector format Features represented in a digital map as points, lines, and areas (polygons). Each individual feature is represented as a single or series of Cartesian coordinates linked to a set of attributes that describe salient characteristics of the feature. *Compare with raster format.*

video frame grabber A circuit board for a computer that converts individual video frames into raster images that can be used by image processing software.

videography The use of videocameras as the primary sensor for generating remote sensing imagery. Videocameras can generate images that are monochrome or multispectral (multiple monochrome cameras with filters or RGB color cameras). Imagery is recorded as a series of videoframes that can be viewed continuously or "grabbed" as individual frames. Videography allows direct audio annotation of key landscape features during data acquisition. With the addition of a digital character generator, a GPS-derived geographic location can be written to each video frame during acquisition.

visible wavelengths The radiation range in which the human eye is sensitive, approximately 0.4–0.7μm.

VISSR Visible Infrared Spin-Scan Radiometer. A sensor carried by the GOES satellites.

volume scattering Scattering of electromagnetic radiation, usually radar, in the interior of a material. This kind of scattering may happen in a vegetation canopy or in the subsurface of soil.

wavelength The reciprocal of the frequency of electromagnetic radiation multiplied by the velocity of light.

References and Further Reading

Achard, F. and F. Blasco. 1990. Analysis of vegetation seasonal evolution and mapping of forest cover in west Africa with the use of NOAA AVHRR HRPT data. *Photogrammetric Engineering and Remote Sensing* 56 (10): 1359–63.

American Society of Photogrammetry. 1980. *Manual of Photogrammetry.* 4th edition. C. C. Slama, C. Theuver, and S. W. Henriksen, eds. Falls Church, Virginia: American Society of Photogrammetry.

Andersen, J. R., E. Hardy, J. Roach, and R. Witmer. 1976. A land use and land cover classification system for use with remote sensor data. *U.S. Geological Survey, Paper* 964.

Bailey, R. G., R. D. Pfister, and J. A. Henderson. 1978. Nature of land and resource classification: A review. *Journal of Forestry* 76: 650–55.

Barber, D. G., P. R. Richard, K. P. Hochheim, and J. Orr. 1991. Calibration of aerial thermal infrared imagery for walrus population assessment. *Arctic* 44: 58–65.

Barnea, D. I. and H. F. Silverman. 1972. A class of algorithm for fast digital image rectification. *IEEE Transactions on Computers* C-21: 179–86.

Barrett, E. C. and L. F. Curtis. 1982. *Introduction to Environmental Remote Sensing.* 2nd edition. London: Chapman and Hall.

Barton, I. J. and J. M. Bathols. 1989. Monitoring floods with AVHRR. *Remote Sensing of Environment* 30:89–94.

Blasco, F., M. F. Bellan, and M. U. Chaudhury. 1992. Estimating the extent of floods in Bangladesh using SPOT data. *Remote Sensing of Environment* 39:167–85.

Bobbe, T., D. Reed, and J. Schramek. 1993. Georeferenced airborne video imagery: Natural resource applications on the Tongass. *Journal of Forestry* 91 (8): 34–37.

Bolton, M. P. and R. L. Specht. 1983. A method for selecting nature conservation reserves. *Australian National Parks and Wildlife Service Occasional Paper* 8:1–32.

Borstad, G. A., D. A. Hill, R. C. Kerr, and B. S. Nakashima. 1992. Direct digital remote sensing of herring schools. *International Journal of Remote Sensing* 13:2191–98.

Brown, A. G., K. J. Gregory, and E. J. Milton. 1987. The use of Landsat multispectral scanner data for the analysis and management of flooding on the river Severn, England, U.K. *Environmental Management* 11:695–702.

Brown, L. R., C. Flavin, and H. Kane. 1992. *Vital Signs*. New York: W. W. Norton.

Chavez, P. S., G. L. Berlin, and L. B. Sowers. 1982. Statistical method for selecting Landsat MSS ratios. *Journal of Applied Photographic Engineering* 8:23–30.

Cliff, A. D. and J. K. Ord. 1973. *Spatial Autocorrelation*. London: Pion Ltd.

Cohen, J. 1960. A coefficient of agreement of nominal scales. *Educational and Psychological Measurement* 20:37–46.

Cowardin, L. M., V. Carter, F. C. Golet, and E. T. Laroe. 1979. *Classification of Wetlands and Deepwater Habitats of the United States*. U.S. Department of Interior Fish and Wildlife Service FWS/OBS-79/31. Washington, D.C.

Cracknell, A. P. and L. W. Hayes. 1990. *Introduction to Remote Sensing*. London: Taylor and Francis.

Cross, A. M. 1992. Monitoring marine oil pollution using AVHRR data observations off the coast of Kuwait and Saudi Arabia during January 1991. *International Journal of Remote Sensing* 13:781–88.

Crumpacker, D. W., S. W. Hodge, D. Friedley, and W. P. Gregg Jr. 1988. A preliminary assessment of the status of major terrestrial and wetland ecosystems on Federal and Indian lands in the United States. *Conservation Biology* 2:103–15.

Curran, P. J. 1985. *Principles of Remote Sensing*. London: Longman.

DeGraaf, R. M. and D. D. Rudis. 1986. New England wildlife habitat, natural history, and distribution. *U.S. Forest Service General Technical Report* NE-108. Washington, D.C.

Delregno, K. J. and S. F. Atkinson. 1988. Nonpoint pollution and watershed management: A remote sensing and geographic information system (GIS) approach. *Lake and Reservoir Management* 4:17–26.

Eidenshink, J. C. 1992. The 1990 conterminous U.S. AVHRR dataset. *Photogrammetric Engineering and Remote Sensing* 58 (6): 809–13.

Everitt, J. H., D. E. Escobar, and F. W. Judd. 1991. Evaluation of airborne video imagery for distinguishing black mangrove (*Avivcennia germinans*) on the lower Texas Gulf Coast. *Journal of Coastal Research* 7 (4): 1169–73.

Everitt, J. H., D. E. Escobar, R. Villarreal, M. A. Alaniz, and M. R. Davis. 1993. Integration of airborne video, global positioning system, and geographic information system technologies for detecting and mapping two woody legumes on rangelands. *Weed Technology* 7:981–87.

Finn, J. T. 1988. Validation of landscape models. In A. Marani, ed., *Advances in Environmental Modelling*, pp. 261–78. Amsterdam: Elsevier.

——. 1993. Use of the average mutual information index in evaluating classification error and consistency. *International Journal of Geographical Information Systems* 7 (4): 349–66.

Friedman, D. E., J. P. Friedel, K. L. Magnussen, R. Kwok, and S. Richardson. 1983. Multiple scene precision rectification of spaceborne imagery with a very few ground control points. *Photogrammetric Engineering and Remote Sensing* 49:1657–67.

Harrison, B. A. and D. L. Judd. 1989. *Introduction to Remotely Sensed Data.* Canberra, Australia: CSIRO (Commonwealth Scientific and Industrial Research Organisation).

Heller, R. C., J. J. Ulliman, R. C. Aldrich, et al. 1983. Forest resource assessments. In R. N. Colwell, ed., *Manual of Remote Sensing*, vol. 2, pp. 2229–324. Falls Church, Virginia: American Society of Photogrammetry.

Herwitz, S. R., D. L. Peterson, and J. R. Eastman. 1990. Thematic mapper detection of changes in the leaf area of closed canopy pine plantations in central Massachusetts. *Remote Sensing of Environment* 29:129–40.

Hill, C. L. 1985. Ducks Unlimited joint research project: Final report. *NASA Report* No. 238. Stennis Space Center.

Hobbs, R. J. and H. A. Mooney. 1990. *Remote Sensing of Biosphere Functioning.* New York: Springer-Verlag.

Hobbs, R. J., J. F. Wallace, and N. A. Campbell. 1989. Classification of native vegetation in the western Australian wheat belt using Landsat MSS data. *Vegetatio* 80:91–105.

Holben, B. N. 1986. Characteristics of maximum-value composite images from temporal AVHRR data. *International Journal of Remote Sensing* 7 (11): 1417–34.

Hurn, J. 1989. GPS: A Guide to the Next Utility. Sunnyvale, Calif.: Trimble Navigation.

Jensen, J. R. 1986. *Introductory Digital Image Processing.* Englewood Cliffs: Prentice-Hall.

Justice, C. O. 1978. The effect of ground conditions on Landsat multispectral scanner data for an area of complex terrain in southern Italy. Unpublished Ph.D. thesis. Reading University, U.K.

Justice, C. O., N. Horning, and N. Laporte. 1993. Remote sensing and GIS contributions to a climage change program in central Africa. In *Central Africa: Global Climate Change and Development*, Technical Report of the Biodiversity Support Program. Washington, D.C.

Justice, C. O. and J. R. G. Townshend. 1981. Integrating ground data with remote sensing. In J. R. G. Townshend, ed., *Terrain Analysis and Remote Sensing*, pp. 38–58. London: Allen and Unwin.

Kauth, R. J. and G. Thomas. 1976. The tasselled cap: A graphical description of the spectral-temporal development of agricultural crops as seen by Landsat. *Proceedings of the Symposium on Machine Processing of Remotely Sensed Data* 4B:41–51.

Koomanoff, V. A. 1989. Analysis of global vegetation patterns: A comparison between remotely sensed data and a conventional map. *Biogeography Research Series Reports* No. 890201. College Park: University of Maryland.

Labovitz, M. L. and E. J. Matsuoko. 1984. The influence of autocorrelation on signature extraction: An example from a geobotanical investigation of Cotter Basin, Montana. *International Journal of Remote Sensing* 5:315–32.

Lillesand, T. M. and R. W. Keifer. 1994. *Remote Sensing and Image Interpretation.* 3rd ed. New York: John Wiley and Sons.

Lo, C. P. 1986. *Applied Remote Sensing.* New York: Longman.

——. 1989. A raster approach to population estimation using high-altitude aerial and space photographs. *Remote Sensing of Environment* 27:59–71.

Loffler, E. and C. Margules. 1980. Wombats *Lasiorhinus latifrons* detected from space. *Remote Sensing of Environment* 9:47–56.

Lubchenco, J., A. M. Olson, L. B. Brubaker, S. R. Carpenter, M. M. Holland, S. P. Hubbell, S. A. Levin, J. A. MacMahon, P. A. Matson, J. M. Melillo, H. A. Mooney, C. H. Peterson, H. R. Pulliam, L. A. Real, P. J. Regal, and P. G. Risser. 1991. The sustainable biosphere initiative: An ecological research agenda. *Ecology* 72 (2): 371–412.

MacConnell, W., J. Stone, D. Goodwin, D. Swartwout, and C. Costello. 1992. Recording wetland delineations on property records: The Massachusetts DEP experience, 1972 to 1992. *Report of the Wetland Mapping Unit, University of Massachusetts.* Amherst, Mass.

Malingreau, J. P. 1977. A proposed land-cover/land-use classification and its use with remote sensing data in Indonesia. *Indonesian Journal of Geography* 7 (33): 5–27.

Marsh, S. E., J. L. Walsh, and C. F. Hutchinson. 1990. Development of an agricultural land-use GIS for Senegal derived from multispectral video and photographic data. *Photogrammetric Engineering and Remote Sensing* 56 (3): 351–57.

Marsh, S. E., J. L. Walsh, and C. Sobrevila. 1994. Evaluation of airborne video data for land-cover classication accuracy assessment in an isolated Brazilian forest. *Remote Sensing of Environment* 48:61–69.

Mather, P. M. 1987. Computer processing of remotely sensed images: An introduction. New York: Wiley.

McDonald, R. A. 1995. Opening the cold war sky to the public: Declassifying satellite reconnaissance imagery. *Photogrammetric Engineering and Remote Sensing* 61 (4): 385–90.

McKendry, G., A. E. Gibson and J. R. Eastman. 1995. Cartographic production. In J. R. Eastman, ed., *IDRISI for Windows User's Guide, Version 1.0,* pp. 11.1–11.18. Worcester, Massachusetts: IDRISI Production, Clark University.

Meisner, D. E. 1986. Fundamentals of airborne video remote sensing. *Remote Sensing of Environment* 19 (1): 63–79.

MicroImages, Inc. 1991. *Soil Map Vectorization by Scanning: Application Note for the Map and Image Processing System.* Lincoln, Nebraska: MicroImages, Inc.

Munsell, 1975. *Munsell Soil Color Charts.* Baltimore: Macbeth Division, Kollmorgen Corp.

Nelson, R., N. Horning, and T. A. Stone. 1987. Determining the rate of forest conversion in Mato Grosso, Brazil, using Landsat MSS and AVHRR data. *International Journal of Remote Sensing* 8:1776–84.

Norwood, V. T. and J. C. Lansing Jr. 1983. Electro-optical imaging sensors. In R. N. Colwell, ed., *Manual of Remote Sensing*, pp. 335–67. Falls Church, Virginia: American Society of Photogrammetry.

Pearce, C. M. 1991. Mapping muskox habitat in the Canadian High Arctic with SPOT satellite data. *Arctic* 44:49–57.

Pressey, R. L. and A. O. Nicholls. 1991. Reserve selection in western division of New South Wales: Development of a new procedure based on land system mapping. In C. R. Margules and M. P. Austin, eds., *Nature Conservation: Cost-Effective Biological Surveys and Data Analysis*, pp. 98–105. East Melbourne, Australia: CSIRO.

Prince, S. D. 1991. Satellite remote sensing of primary production: Comparison of results for Sahelian grasslands, 1981–1988. *International Journal of Remote Sensing* 12:1313–30.

Prince, S. D., C. O. Justice, and S. O. Los. 1990. *Remote Sensing of the Sahelian Environment: A Review of the Current Status and Future Prospects.* Bruxelles: Van Ruys.

Quattrochi, D. A. and R. E. Pelletier. 1990. Remote sensing for analysis of landscapes: An introduction. In M. G. Turner and R. H. Gardner, eds., *Quantitative Methods in Landscape Ecology: The Analysis and Interpretation of Landscape Heterogeneity*, pp. 51–76. New York: Springer-Verlag.

Richards, J. A. 1984. Thematic mapping from multitemporal image data using the principal components transformation. *Remote Sensing of Environment* 16:35–46.

——. 1986. *Remote Sensing Digital Image Analysis: An Introduction.* New York: Springer-Verlag.

Rosenfield, G. H. and K. Fitzpatrick-Lins. 1986. A coefficient of agreement as a measure of thematic classification accuracy. *Photogrammetric Engineering and Remote Sensing* 52:223–27.

Sabins, F. F. 1986. *Remote Sensing.* 2nd ed. San Francisco: Freeman.

Sader, S. A. and J. C. Winne. 1992. RGB-NDVI color composites for visualizing forest change dynamics. *International Journal of Remote Sensing* 13 (16): 3055–68.

Schwaller, M. R., C. E. Olson Jr., Z. Ma, Z. Zhu, and P. Dahmer. 1989. A remote sensing analysis of Adelie penguin rookeries. *Remote Sensing of Environment* 28:199–206.

Scott, J. M., F. Davis, B. Csuti, R. Noss, B. Butterfield, C. Groves, H. Anderson, S. Caicco, F. D'Erchia, T. C. Edwards Jr., J. Ulliman, and R. G. Wright. 1993. Gap analysis: A geographic approach to protection of biological diversity. *Wildlife Monographs* 123:1–41.

Sellers, P. J., S. I. Rasodand H.-J. Bolle. 1990. A review of satellite data algorithms for studies of the land surface. *Bulletin of the American Meteorological Society* 717 (10): 1429–47.

Sidle, J. G., D. E. Carlson, E. M. Kirsch, and J. J. Dinan. 1992. Flooding mortality and habitat renewal for least terns and piping plovers. *Colonial Waterbirds* 15:132–36.

Sidle, J. G., H. G. Nagel, R. Clark, C. Gilbert, D. Stuart, K. Willburn, and M. Orr. 1993. Aerial thermal infrared imaging of sandhill cranes on the Platte River, Nebraska. *Remote Sensing of Environment* 43:333–41.

Sidle, J. G. and J. W. Ziewitz. 1990. Use of aerial videography in wildlife habitat studies. *Wildlife Society Bulletin* 18:56–62.

Skrdla, M. P. 1992a. *A Guide to Map and Image Processing: Reference Manual for the Map and Image Processing System.* Lincoln, Nebraska: MicroImages, Inc.

——. 1992b. *Map and Poster Layout: Application Note for the Map and Image Processing System.* Lincoln, Nebraska: MicroImages, Inc.

——. 1993. *Feature Mapping: Application Note for the Map and Image Processing System.* Lincoln, Nebraska: MicroImages, Inc.

Slater, P. N., F. J. Doyle, L. F. Norman, and R. Welch. 1983. Photographic systems for remote sensing. In R. N. Colwell, ed., *Manual of Remote Sensing*, pp. 231–91. Falls Church, Virginia: American Society of Photogrammetry.

Specht, R. L. 1975. The report and its recommendations. In F. Fenner, ed., *A National System of Ecological Reserves in Australia*, pp. 11–16. Canberra: Australian Academy of Science.

Spitzer, D., R. Laane, and J. N. Roosekrans. 1990. Pollution monitoring of the North Sea using NOAA AVHRR imagery. *International Journal of Remote Sensing* 11 (6): 967–78.

Stone, T. A., I. F. Brown, and G. M. Woodwell. 1991. Estimation by remote sensing of deforestation in central Rondonia, Brazil. *Forest Ecology and Management* 38 (3–4): 291–304.

Strahler, A. H., J. E. Estes, P. F. Maynard, F. C. Mertz, and D. A. Stow. 1980. Incorporating collateral data in Landsat classification and modeling procedures. *Proceedings 14th International Symposium on Remote Sensing of Environment*, pp. 927–942.

Stumpf, R. P. and M. A. Tyler. 1988. Satellite detection of bloom and pigment distributions in estuaries. *Remote Sensing of Environment* 24:385–404.

Tou, J. T. and R. C. Gonzalez. 1974. *Pattern Recognition Principles.* New York: Addison-Wesley.

Townshend, J. R. G. 1981. *Terrain Analysis and Remote Sensing.* London: Allen and Unwin.

TREES. 1991. Tropical Ecosystem Environment Observations by Satellites: Strategy proposal, 1991–93. Part 1: AVHRR data collection and analysis. *TREES Series A Technical Document*, No. 1 (EUR 14026 EN). Ispra, Italy: JRC/ESA.

Tucker, C. J., W. W. Newcomb, S. O. Los, and S. D. Prince. 1991. Mean and inter-year variation in growing season normalized difference vegetation index for the Sahel, 1981–1989. *International Journal of Remote Sensing* 12 (6): 1133–36.

Tucker, C. J., J. R. G. Townshend, and T. E. Gott. 1985. African land-cover classification using satellite data. *Science* 227 (4685): 369–75.

Tucker, C. J., C. L. VanPraet, M. J. Sharman, and G. Van Ittersum. 1985. Satellite remote sensing of total herbaceous biomass production in the Senegalese Sahel, 1980–1984. *Remote Sensing of Environment* 17:233–49.

Tyler, M. A. and R. P. Stumpf. 1989. Feasibility of using satellites for detection of kinetics of small phytoplankton blooms in estuaries: Tidal and migrational effects. *Remote Sensing of Environment* 27:233–50.

Ulanowicz, R. E. 1986. *Growth and Development*. New York: Springer-Verlag.

Vitousek, P. M., P. R. Ehrlich, A. H. Ehrlich, and P. M. Matson. 1986. Human appropriation of the products of photosynthesis. *Bioscience* 36 (6): 368–73.

Western, D. and M. Pearl, eds. 1989. *Conservation Biology for the Next Century*. New York: Oxford University Press.

Wilkie, D. S. 1989. Performance of a backpack GPS in a tropical rain forest. *Photogrammetric Engineering and Remote Sensing* 55 (12): 1747–49.

Wilson, E. O. 1992. *The Diversity of Life*. Cambridge, Mass.: Harvard University Press.

Woodcock, C. E. and A. H. Strahler. 1987. The factor of scale in remote sensing. *Remote Sensing of Environment* 21:311–32.

Wright, R. 1993. Airborne videography: Principles and practice. *Photogrammetric Record* 14 (81): 447–57.

Zhu, Z. and D. L. Evans. 1992. Mapping midsouth forest distributions. *Journal of Forestry* 90 (12): 27–30.

Index

Italicized numerals refer to illustrations or tables on that page. Bold numerals indicate that a term is defined on that page.